AF905414

Large Dams in Asia

Advances in Asian Human-Environmental Research

Series Editor

Prof. Marcus Nüsser
South Asia Institute, University of Heidelberg, Germany

Editorial Board

Prof. Eckart Ehlers, University of Bonn, Germany
Prof. Harjit Singh, Jawaharlal Nehru University, New Delhi, India
Prof. Hermann Kreutzmann, Freie Universität Berlin, Germany
Prof. Ken Hewitt, Waterloo University, Canada
Prof. Urs Wiesmann, University of Bern, Switzerland
Prof. Sarah J. Halvorson, University of Montana, USA
Dr. Daanish Mustafa, King's College London, UK

Aims and Scope

The series aims at fostering the discussion on the complex relationships between physical landscapes, natural resources, and their modification by human land use in various environments of Asia. It is widely acknowledged that human-environment-interactions become increasingly important in area studies and development research, taking into account regional differences as well as bio-physical, socio-economic and cultural particularities.

The book series seeks to explore theoretic and conceptual reflection on dynamic human-environment systems applying advanced methodology and innovative research perspectives. The main themes of the series cover urban and rural landscapes in Asia. Examples include topics such as land and forest degradation, glaciers in Asia, mountain environments, dams in Asia, medical geography, vulnerability and mitigation strategies, natural hazards and risk management concepts, environmental change, impacts studies and consequences for local communities. The relevant themes of the series are mainly focused on geographical research perspectives of area studies, however there is scope for interdisciplinary contributions.

For further volumes:
http://www.springer.com/series/8560

Marcus Nüsser
Editor

Large Dams in Asia

Contested Environments between Technological Hydroscapes and Social Resistance

Springer

Editor
Marcus Nüsser
Department of Geography
South Asia Institute
Heidelberg University
Heidelberg, Germany

ISSN 1879-7180 ISSN 1879-7199 (electronic)
ISBN 978-94-007-2797-7 ISBN 978-94-007-2798-4 (eBook)
DOI 10.1007/978-94-007-2798-4
Springer Dordrecht Heidelberg New York London

Library of Congress Control Number: 2013952931

© Springer Science+Business Media Dordrecht 2014
This work is subject to copyright. All rights are reserved by the Publisher, whether the whole or part of the material is concerned, specifically the rights of translation, reprinting, reuse of illustrations, recitation, broadcasting, reproduction on microfilms or in any other physical way, and transmission or information storage and retrieval, electronic adaptation, computer software, or by similar or dissimilar methodology now known or hereafter developed. Exempted from this legal reservation are brief excerpts in connection with reviews or scholarly analysis or material supplied specifically for the purpose of being entered and executed on a computer system, for exclusive use by the purchaser of the work. Duplication of this publication or parts thereof is permitted only under the provisions of the Copyright Law of the Publisher's location, in its current version, and permission for use must always be obtained from Springer. Permissions for use may be obtained through RightsLink at the Copyright Clearance Center. Violations are liable to prosecution under the respective Copyright Law.
The use of general descriptive names, registered names, trademarks, service marks, etc. in this publication does not imply, even in the absence of a specific statement, that such names are exempt from the relevant protective laws and regulations and therefore free for general use.
While the advice and information in this book are believed to be true and accurate at the date of publication, neither the authors nor the editors nor the publisher can accept any legal responsibility for any errors or omissions that may be made. The publisher makes no warranty, express or implied, with respect to the material contained herein.

Cover image: Nomads near Nanga Parbat, 1995. Copyright © Marcus Nüsser (used with permission)

Printed on acid-free paper

Springer is part of Springer Science+Business Media (www.springer.com)

Preface

Based on the underlying paradigms of modernisation and the concrete idea of implementation of large infrastructure projects, the quest for national development and improvement of local livelihoods has generated a strong boost for the demand of large dams in the Global South, especially in Asia. The belief in the panacea that gargantuan hydro-projects offer for the betterment of nations and peoples has run roughshod over other environmental and social concerns. Whereas these large-scale transformations of fluvial environments into technological hydroscapes serve to provide hydropower, flood control, and water supply for irrigation and industrial and urban uses, their massive adverse effects have evoked controversies of developmental and environmental impacts.

The contributions in this edited volume explore the various dimensions of the large dams controversy in Asia from a critical perspective. Most of these contributions originate from the research project 'Large dams: Contested environments between hydro-power and resistance', which ran from 2008 to 2011 as part of the Cluster of Excellence 'Asia and Europe in a Global Context: Shifting Asymmetries in Cultural Flows', Heidelberg University. I am grateful to the German Research Council (DFG) and the German Council of Science and Humanities (Wissenschaftsrat) for funding. The continuous support of the directorate and administrative staff of the cluster is gratefully acknowledged. I am indebted to Thomas Lennartz (Heidelberg), who worked hard to standardize formats of all individual contributions. It is hoped that this volume will be beneficial to those looking to gain an overview of the large dams debate. At the same time, the individual chapters may offer insights from case studies that should be useful to a specialist audience.

Department of Geography
South Asia Institute, Heidelberg University
Heidelberg, Germany

Marcus Nüsser

Contents

Contributors

Ravi Baghel Department of Geography, South Asia Institute, Heidelberg University, Heidelberg, Germany

Kumkum Bhattacharyya Department of Geography, University of Central Arkansas, Conway, AR, USA

Nirmalya Choudhury Senior Research Associate, Council on Energy Environment and Water (CEEW), New Delhi, India

Rohan D'Souza Centre for Studies in Science, Policy School of Social Sciences, Jawaharlal Nehru University, New Delhi, India

Alexander Erlewein Department of Geography, South Asia Institute, Heidelberg University, Heidelberg, Germany

Marcus Nüsser Department of Geography, South Asia Institute, Heidelberg University, Heidelberg, Germany

Miriam Seeger Institute of Chinese Studies, Heidelberg University, Heidelberg, Germany

Hanna Werner Institut für Asien- und Afrikawissenschaften, Seminar für Südasien-Studien, Humboldt-Universität zu Berlin, Berlin, Germany

Michael J. Wiley School of Natural Resources and Environment, University of Michigan, Ann Arbor, MI, USA

Abbreviations

ADHPL	Allain Duhangan Hydro Power Limited
BCE	before the Common Era
BNP	Bhakra Nangal Project
CAS	Chinese Academy of Sciences
CASS	Chinese Academy of Social Sciences
CB ratio	cost-benefit ratio
CCP CC	Chinese Communist Party Central Committee
CDM	Clean Development Mechanism
CE	Common Era
CER	Certified Emission Reduction
CO_2	carbon dioxide
CO_2e	carbon dioxide equivalent
CS	Cadastral Survey
CV	coefficient of variation
DNA	Designated National Authority
DOE	Designated Operational Entities
DVC	Damodar Valley Corporation
EAC	Expert Appraisal Committee
EF	exceedance frequency
EFR	environmental flow requirement
EIA	environmental impact assessment
ENSO	El Niño/La Niña-Southern Oscillation
ET	emission trading
GIS	geographic information systems
GHG	greenhouse gas
GoS	Geography of Science
GUP	Government of Uttar Pradesh
GWIL	Gujarat Water Infrastructure Limited
HEC	Hydrologic Engineering Center
HFC	hydrofluorocarbon
IRS	Indian Remote Sensing

ICOLD	International Commission on Large Dams
IFC	International Finance Corporation
JI	joint implementation
KWDT	Krishna Water Disputes Tribunal
LISS	Linear Imaging Self-Scanning
MoEF	Ministry of Environment and Forest, Government of India
MPRVD	Multi-purpose River Valley Development
MRO	manager reservoir operation
MW	megawatt
N_2O	nitrous oxide
NIR	near infrared
NGO	non-governmental organisation
NRSA	National Remote Sensing Agency
P	precipitation
PAP	Project Affected Person
PDD	Project Design Document
PFC	perfluorocarbon
Q	hydrologic discharge
R&R	resettlement and rehabilitation
RBO	River Basin Organisation
RS	Revision Survey
SANDRP	South Asian Network on Dams, Rivers and People
SD	standard deviation
SoI	Survey of India
SSP	Sardar Sarovar Project
SSK	Sociology of Scientific Knowledge
STS	Science and Technology Studies
SWIR	short-wave infrared
TBVSS	Tehri Bandh Virodhi Sangharsh Samiti (Committee to Oppose the Tehri Dam)
THDC	Tehri Hydro Development Corporation
TINA	There Is No Alternative
TVA	Tennessee Valley Authority
UNFCCC	United Nations Framework Convention on Climate Change
WAPDA	Water and Power Development Authority (Pakistan)
WCD	World Commission on Dams
WFD	Water Framework Directive
WWF	World Wildlife Fund

Units of Measure

ft^3/s	cubic feet per second (cusec)	1 $ft^3/s \approx 0.028\ m^3/s$
GW	gigawatt	1 GW = 1,000,000,000 W
ha	hectares	1 ha = 10,000 m^2
kV	kilovolt	1 kV = 1,000 V
maf	million acre feet	1 acre foot ≈ 1233.5 m^3
mha	million hectares	
MW	megawatt	1 MW = 1,000,000 W

Chapter 1
Technological Hydroscapes in Asia: The Large Dams Debate Reconsidered

Marcus Nüsser

Abstract Large dams have evoked contentious debates about the trajectories and effectiveness of development, environmental impacts and social justice. As is especially the case in the countries of the Global South, particularly in Asia, the construction and operation of large dams are amongst the most prestigious but also most sensitive development issues. These large-scale transformations of fluvial environments into technological hydroscapes serve to provide hydropower, flood control and water supply for irrigation, industrial and urban uses. In turn, negative socioeconomic consequences through displacement, resettlement and insufficient compensation as well as adverse environmental impacts emerge. From a political ecology perspective, these massive development interventions can be seen as contested environments shaped by distinct actor constellations and the underlying aspirations of modernisation and social resistance. This introductory chapter explores the various dimensions of the large dams controversy in Asia and retraces the historical course of the debate in the context of changing development paradigms, underlying power asymmetries and the social constructions of nature.

Keywords Large dams • Hydroscapes • Socio-hydrology • Development • Politicised environments

1.1 Framing the Large Dams Debate

Being among the most massive infrastructure projects built worldwide, large dams provide hydroelectric power, flood control and water supply for irrigation, industrial and urban uses. Large dams constitute powerful symbols of modernisation, national

M. Nüsser (✉)
Department of Geography, South Asia Institute, Heidelberg University, Im Neuenheimer Feld 330, 69120 Heidelberg, Germany
e-mail: marcus.nuesser@uni-heidelberg.de

M. Nüsser (ed.), *Large Dams in Asia*, Advances in Asian Human-Environmental Research, DOI 10.1007/978-94-007-2798-4_1,
© Springer Science+Business Media Dordrecht 2014

prestige and human dominance over nature. At the same time, these gigantic 'tools' for the management of water resources have evoked controversial debates on development paradigms, development effectiveness, social justice and sustainability (e.g. McCully 2001; Khagram 2004; Kreutzmann 2004; D'Souza 2006). The economic benefits of dam building for hydropower generation and effective water management for irrigated agriculture, drinking water supply and flood control are almost always in contrast to the adverse consequences for local populations and negative environmental impacts. Among the most common socioeconomic concerns are insufficient compensation for displaced people and the related lack of long-term development perspectives. Against this backdrop, the main arguments in the large dams controversy are rooted in the classical development theories of modernisation and dependency. Large dams are textbook examples that highlight some of the most sensitive and contested development issues as well as complex socio-hydrological interactions. Although a huge body of literature on large dams and river control has been published over the past decades and the principal arguments are repeated in the context of various case studies, the transformation of fluvial environments are seldom framed within a larger picture of contested development paradigms across changing historical and political contexts.

According to the inventory of the World Commission on Dams (WCD) in 2000, there are more than 45,000 large dams,[1] all but 5,000 of them built since 1950. Most of these dams were constructed in Asia, with China and India being among the most prolific dam-building countries in the world (Gleick 1998, p. 78; McCully 2001, p. 66). China has by far the world's largest installed hydropower capacity, with aspirations for further expansion (Gleick 2012, p. 129). Moreover, Chinese funding institutions and engineering companies are increasingly involved in dam projects in neighbouring and overseas countries (Brewer 2007; Gleick 2012, p. 308). China's involvement in the construction of the Diamer Basha Dam in the Indus Valley of northern Pakistan, with the promise of several thousand experienced workers from the Three Gorges Dam, is just one example. This dimension of technology and knowledge transfer is in many ways similar to the expertise flowing from the Tennessee Valley Authority (TVA) in the USA to India in the 1950s. In the Himalayan region, the governments of India, Pakistan, Nepal and Bhutan are transforming the upper reaches of the mountain drainage system into the subcontinental powerhouse of South Asia at an unprecedented pace. In the same direction and at an even faster rate, China taps the water resources of the Tibetan

[1] In order to define large dams, the *International Commission on Large Dams* (ICOLD), a professional organisation of the dam-building industry, established in 1928, offers a set of criteria. According to their well-established definition, a large dam is one whose height exceeds 15 m or whose height is between 10 and 15 m, if it meets at least one of the following conditions: the crest length of the dam is not less than 500 m, the spillway discharge potential exceeds 2,000 m^3 per second or the reservoir volume is not less than 1 million m^3. A major dam is defined by ICOLD as a dam meeting at least one of the following four requirements: the dam is at least 150 m high, the dam volume exceeds 15 million m^3, the reservoir storage capacity exceeds 25 billion m^3 or the installed electrical generation capacity is at least 1,000 MW.

Photo 1.1 The controversy of the Yamdrok Yumtso Lake. This unique freshwater lake, located at an altitude of 4,400 m on the Tibetan plateau, is tapped to power a 90 MW electric generator (Photograph: Marcus Nüsser, 12 May 2002)

plateau to cope with growing water and energy demands in the region (Photo 1.1) and in the urban agglomerations of China. The importance of the Asian high mountain belts as sources of freshwater supply for the adjoining lowlands has led to them being characterised as 'water towers' (Viviroli et al. 2007; Viviroli and Weingartner 2008). Apart from this fundamental hydrological attribute, the Himalayan region, including the Tibetan plateau, increasingly gains the additional function of a 'power tower', intensifying and diversifying the resource transfer from the mountains to the plains of the subcontinent (Erlewein and Nüsser 2011, p. 302; Erlewein 2012, p. 33). Most prominent examples are the northern states of India, namely, Himachal Pradesh, Uttarakhand, Sikkim (Photo 1.2) and Assam, which supply the economic centres in the adjoining lowlands with hydroelectric energy. In addition, India supports dam building in the upper riparian countries of Bhutan and Nepal so as to fulfil the nations' energy demands. All of these South Asian countries are affected by chronic energy deficiencies, resulting in load shedding to various degrees. The precarious energy shortage became most obvious when a huge electricity blackout, affecting 1.2 million people, occurred in India in July 2012 and gained international media attention. To date, the reality remains the same for many people along the Himalayan arc, from Pakistan to Nepal, who only receive a few hours of power supply on a rotational basis. Against this background and the states' economic development aspirations, the government's perspectives are to increase the nation's hydropower capacities for domestic and industrial use. The same holds

Photo 1.2 Damming Sikkim: A cascade of 36 hydropower plants are planned along the Teesta River in one of India's Himalayan states (Photograph: Marcus Nüsser, 9 April 2011)

true for the expansion of irrigation systems so as to improve food security and drinking water supplies. Juxtaposed to these potential benefits of dam building is a swath of negative externalities.

Worldwide, 40–80 million people have been displaced due to dam construction and related flooding (World Commission on Dams 2000, p. 16). For the Three Gorges Dam in China alone, almost 1.2 million persons were displaced (Li et al. 2001, p. 201; Yardley 2007). The affected people, sometimes referred to as 'oustees', who are mostly excluded from planning processes, are then confronted with an endangered livelihood security, which is further compounded by inadequate compensation for their material and cultural losses. Whereas local residents are considered 'refugees in an unacknowledged war' (Roy 2001, p. 65) by environmental and social activist groups, nation-states identify them as beneficiaries of the prospective improvements. Struggles over power between nation-states, international funding agencies and industries, non-governmental organisations (NGOs) and adversely affected persons testify to a 'politicised environment' in the sense of Bryant and Bailey (1997, p. 28). Central to this concept is the recognition that environmental conflicts cannot be understood without taking into consideration the political and socioeconomic contexts within which they emerge. Apparently, arguments of actors on the national level are linked to economic interests within and across countries and influenced by the international dam-building industry and multilateral funding institutions. Both advocates and opponents of large hydro-projects form coalitions to

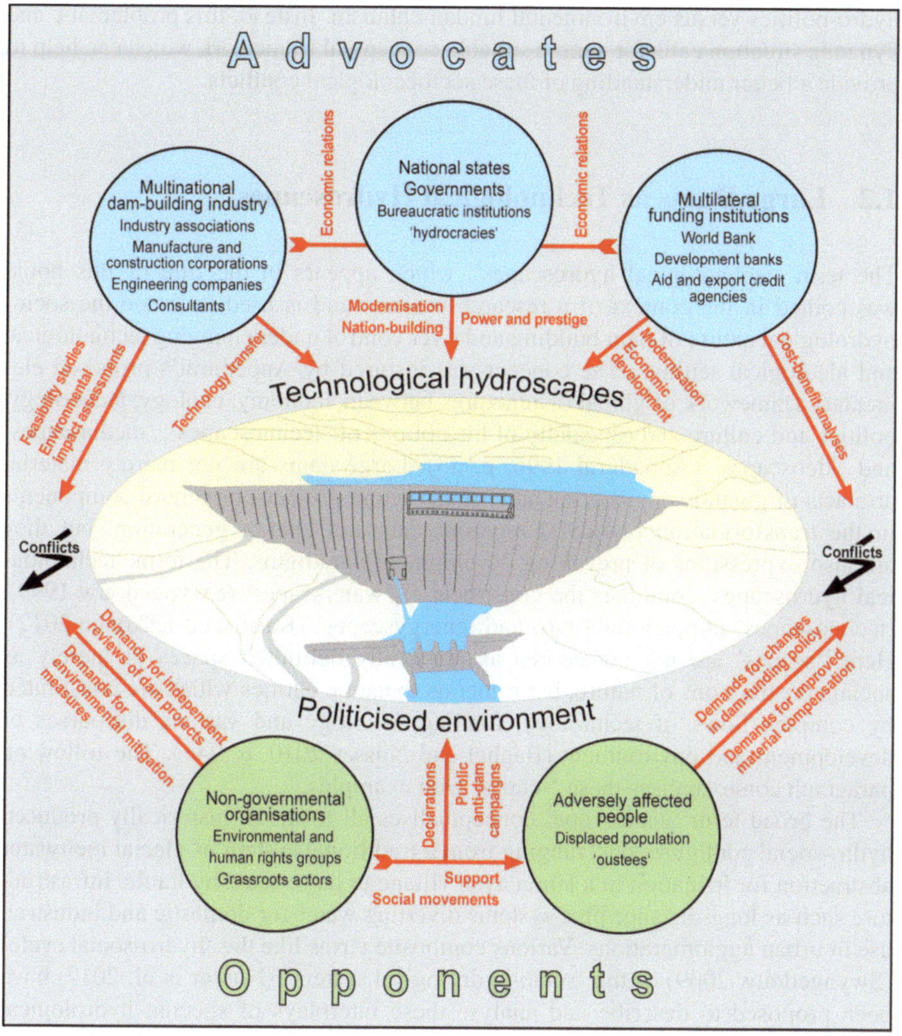

Fig. 1.1 The politicised environment and the principal actor constellation in the large dams debate (Modified after Nüsser 2003, p. 22)

strengthen their position during the planning and implementation phase. The actor-orientated approach of political ecology provides an ideal framework to identify the character, scope and dynamics of these contested politicised environments, where place-based and non-place-based actors are involved (Nüsser 2003; Baghel and Nüsser 2010, Fig. 1.1). Moreover, changing social constructions of the role of the environment in fostering development shape struggles over power, interests and institutions (Adger et al. 2001; Watts and Peet 2004; Agrawal 2005). The deep-rooted differences between various actor groups cannot be simplified as modernised hydro-politics versus environmental fundamentalism. Instead, this problematic and dynamic situation calls for a more suitable conceptual framework which can help to provide a better understanding of these socioecological conflicts.

1.2 Large Dams as Technological Hydroscapes

The term 'technological hydroscapes', which appears in the title of this book, was coined in the context of a research project[2] and is used to frame the socio-hydrological nature of dam building and river control under changing technological and ideological settings. The concept was inspired by Appadurai's proposed elementary framework of global disjunctures between economy, ecology, technology, politics and culture and especially of his notions of 'technoscapes', 'mediascapes' and 'ideoscapes' (Appadurai 1996, p. 33). Large dams are not merely material artefacts of gigantic engineering and infrastructure projects, or central components in the transformation of fluvial environments and energy generation, but they are also expressions of prevailing development paradigms. The term 'technological hydroscapes' combines the constructs of 'waterscapes' (Swyngedouw 1999), 'technoscapes' (Appadurai 1996) and 'energyscapes' (Kaisti and Käkönen 2012). Here, 'scapes' are not considered as physically delimited spaces or merely as social constructions of nature, but rather as dynamic entities which are constituted by complex flows of technology, funding, ideology and various discourses of development and environment (Baghel and Nüsser 2010, p. 241). The following paragraph contextualises these 'scapes' with examples.

The broad term 'waterscape' conceptualises all forms of historically produced hydro-social configurations, ranging from a traditional system of glacial meltwater abstraction for irrigation in a Himalayan village to large-scale hydraulic infrastructure such as long-distance pipe systems diverting water for domestic and industrial use in urban agglomerations. Various composite terms like the 'hydro-social cycle' (Swyngedouw 2009) or the 'socio-hydrological system' (Nüsser et al. 2012) have been proposed to describe and analyse these interplays of specific hydrological conditions and dynamics, socioeconomic development processes and institutional arrangements. In the context of large dams, the technological dimension of these socio-hydrological interventions is of utmost importance and leads to the use of the term 'technological hydroscape'. Dam building and river embankments evolved with improvements in engineering knowledge, construction skills and progress in hydrological analyses. Current examples of technological progress are the run-of-river power plants. These designs require smaller reservoirs, from which water is diverted into head race tunnels before reappearing some kilometres down the valley,

[2]The research project 'Large dams: Contested environments between hydro-power and resistance' was funded from 2008 to 2011 as part of the Cluster of Excellence 'Asia and Europe in a global context: Shifting asymmetries in cultural flows', Heidelberg University.

where the powerhouses are located, and with an increased hydraulic head. The design is widely used in the Himalayan states of India, often in the form of a series of cascades and turbines. Although such systems are socially more acceptable, due to the smaller areas of inundation, there are adverse ecological consequences like river fragmentation, leading to degradation of habitat conditions for several fish and plant species (Erlewein 2012, p. 30). As many of these projects will exclusively be for hydropower generation, their construction will result in an uneven distribution of potential gains and losses, which in turn 'accounts for a serious legitimacy deficit in India's ambitious hydropower development plans in the region' (Baruah 2012, p. 41). The term 'energyscape' refers to the spatial and temporal dimensions of energy security, energy access and the corresponding infrastructure such as concrete artefacts, power houses and power grids (Kaisti and Käkönen 2012). The above-mentioned Himalayan 'power tower' may serve as an example to illustrate the spatial dimension of massive energy transfer from the mountain belt to the plains as well as across national borders. The integration of these different dimensions ('scapes') justifies the use of the composite term 'technological hydroscapes'. This term captures not only the physical transformation of the landscape but also the implementation of new water and energy governance systems. The fundamental shift in the rationale and scale from local water use to the human domination of water systems was only made possible with the advancement of modern technology.

These emergent systems of water control were accompanied by a wide constellation of actors, for which Molle et al. (2009) coined the term 'hydrocracies' as an abbreviation of hydraulic bureaucracies. The various stakeholders consist of influential elites of technical engineers, hydrologists, political and financial leaders. This powerful actor constellation reinforced centralised planning and management of river control in relation to the making of modern nation-states. The rationale for the growth of the bureaucratic apparatus was based on the logic of further expansion of river control and complete transformation of fluvial environments. This logic also propagated the foundation and growth of multiple statal organisations, each competing for influence and reasoning for their own existence. One example is India's 'Ministry of Water Resources', which has a complicated history, after having variously been a 'Department of Irrigation' under Ministries associated with Irrigation, Power, Mines, Scientific Research and Agriculture in different periods until finally becoming a Ministry of its own in 1985 (Ministry of Water Resources 2012). Pakistan's Water and Power Development Authority (WAPDA), established in 1958, is another case in point. This centralised government-owned organisation was founded for the purpose of coordinating and giving a unified direction to the development of water and power schemes, which were previously being dealt with by respective electricity and irrigation departments at the provincial level. The emergence of such powerful and centralised parastatal agencies (hydrocracies) is by no means limited to the Asian context, as the example of the Lesotho Highlands Development Authority in southern Africa shows (Nüsser 2003). Apparently, these bureaucratic institutions have a vital self-interest to continue in the dam-building business. These examples suggest how large dams have served changing bureaucratic structures and development priorities. Departing from the discussion of the more generalised term 'technological hydroscapes', this introductory chapter retraces the historical course of the large dams debate in the context of changing development paradigms and underlying power asymmetries.

1.3 Ideological Continuities and Disjunctures

The principal relations between river control, water regulation and distribution measures and the formation of power structures in so-called 'hydraulic societies' were first discussed in the classical study by Wittfogel (1957). Despite the long history of water management in human civilisation, through the construction of dams, reservoirs and embankments, extensive implementation of these technologies did not commence until the middle of the twentieth century with the notable exception of extensive river control measures and dam building in Germany before the 1930s (Blackbourn 2006, pp. 189–249). However, the German expertise in dam construction was largely domestic. The contemporary large dam era started in the USA with the construction of the Hoover Dam in the 1930s. Subsequently, many large dams were built in the Soviet Union, following Stalin's concept of a 'transformation of nature into a machine for the communist state' (McCully 2001, p. 240; Molle et al. 2009, p. 334). After the period of decolonisation, widespread construction of large dams started in the countries of the Global South. Examples range from Ghana's Akosombo Dam (completed 1965), which resulted in the formation of Lake Volta, to the Aswan Dam in southern Egypt (completed 1970) and the Bhakra Dam in northern India (completed 1963). Since then, China and India have become the most prominent dam-building nations. In the course of this 'hydraulic mission' (Molle 2009, p. 490) to control river flow for 'the common good', so that 'not a single drop of water' reaches the sea before being used by Man (Molle et al. 2009, p. 332), large dams have come to symbolise this human dominance over nature (Photo 1.3). Additionally, they form icons of modernity and expressions of national prestige and emancipation from colonial rule. As the first Indian Prime Minister, Jawaharlal Nehru, stated, large dams are 'temples of resurgent India' and 'symbols of India's progress'. The same ideology of modernisation was repeated during Mao Zedong's 'Great Leap Forward' resulting in the construction of more than 600 large dams per year (Gleick 1998, p. 78, McCully 2001, p. 19). These efforts of river control were based not only on development aspirations but also upon the belief of the need to conquer nature as a threat, or as Kaika (2006, p. 276) succinctly put it as the 'modernist quest to tame, control, and discipline nature'.

These symbols of modernisation are not solely for the internal benefit and prestige of nation-states, but they also have strategic geopolitical implications. India's Bhakra Dam, located close to the border of the Indian states Himachal Pradesh and Punjab on the Sutlej River, which eventually flows into Pakistan, is a case in point. The Bhakra Dam was planned and initiated under the British administration and was completed by Indian engineers under American supervision. Its power station on the right bank was built and later upgraded with Soviet

Photo 1.3 A cement company's advertisement in the Indian Punjab: a motif of human domination over nature (Photograph: Marcus Nüsser, 30 September 2009)

assistance and technology during the height of the Cold War. India's control of the water flow through the dam was a major source of contestation from the Pakistan side of the Punjab owing to the need for regular water supply for irrigation during the green revolution in both countries. This eventually led to the signing of the Indus Water Treaty in 1960 (Baghel 2013, pp. 11, 35). Generally, such large hydropower projects include a flow of international expertise and financial investment, setting the blueprints for future development paths. In the context of China, the technological capacities needed for construction and management of hydrological infrastructure was initially provided by the Soviet Union, with thousands of engineers being trained. Regardless of the political and ideological positions of respective countries, large dams were considered as important foundations for national development, legitimised by the paradigms of modernisation theory.

A second phase in the debate started in the late 1980s and was marked by a paradigm shift in the perception of the values of large dams, related to the massive adverse socioeconomic and environmental consequences. Issues of displacement and insufficient remuneration of local populations came to the fore in the public discourse. In this context, mega-projects such as China's Three Gorges Dam or India's Narmada Cascade are prominent examples of development-induced displacement, dispossession and the emergence of resistance to such projects and their underlying ideologies. This resistance is not restricted to local activism of adversely affected people, but rather is integrated in the practice of international networks of environmental and human rights movements, supported by civil society actors (Florini 2000). Globally operating organisations like International Rivers occupy a key position in this regard, successfully raising awareness of the adverse effects of dams and actively supporting grassroots movements. In India, the Narmada Bachao Andolan movement occupied a key role in the discourse (Khagram 2004; Nilsen 2010) and launched powerful opposition campaigns, inspiring other resistance movements within and outside the country. Prominent personalities such as Arundhati Roy and Medha Patkar gained global recognition for their active role in speaking out for the voiceless, as Roy exclaimed:

> Big Dams started well, but have ended badly. There was a time when everybody loved them, everybody had them – the Communists, Capitalists, Christians, Muslims, Hindus, Buddhists. There was a time when Big Dams moved men to poetry. Not any longer. … Big Dams are obsolete. They're uncool. They're undemocratic. … They're brazen means of taking water, land and irrigation away from the poor and gifting it to the rich. Their reservoirs displace huge populations of people, leaving them homeless and destitute (Roy 2001, p. 57).

As opposed to the Bhakra Dam, which gained widespread support as an icon for economic growth, the Narmada dam cascade became a symbol of social inequity, resistance and an expression of failed development (Gadgil and Guha 1994; Drèze et al. 1997; Dittrich 2004). Besides the usual socioeconomic and environmental concerns, the Tehri Dam (completed in 2006) in the Himalayan state of Uttarakhand, being the highest dam in India (Photo 1.4), has been criticised for the risk associated with it being located in a seismically active zone. The opponents to such projects sum up their adverse consequences as 'dammed development'. Whereas the debate in India takes place in a democratic setting with multiple actors voicing their opinions (all too often with little real effect), the decision-making in China is primarily taken by the Communist Party of China. Despite international concerns, dam building there continues unabated, with little local opposition. The gargantuan Three Gorges Dam, a dream of Mao Zedong, finally came to completion in 2006. This dam has been reported to be the world's largest dam, biggest power plant and biggest consumer of building material. Moreover, it has created the largest number of reservoir 'displacees' as a special group of forced migrants (Li et al. 2001, p. 201; Yardley 2007). One notable opponent to its construction was Dai Qing, who went to prison in 1989 for her critical work on the massive project (Dai 1994, 1998).

From the early enthusiasm surrounding the emergence of technological hydroscapes as economic drivers of modernisation to the disillusion caused by the negative social and environmental outcome, the large dams debate has been tumultuous. It seems that from once being praised as temples, they have now turned into tombs of concrete (Khagram 2003). Currently, the arguments of dam opponents contain key terms like marginalisation, asymmetrical power relations, uneven development, centre-periphery dichotomies, structural inequalities and the like, resembling the classical rhetoric of dependency theory.

Despite ongoing controversy over large dam projects, the discussion at the turn of the century was characterised by efforts to look for suitable compromises that meet the requirements of different development perspectives (Iyer 2003, 2007). In this

Photo 1.4 Tehri Dam, a rock- and earth-fill embankment dam, at 260 m the highest in India (Photograph: Marcus Nüsser, 15 September 2008)

context, the World Commission on Dams (WCD) was founded in 1997 to elaborate common ground in the negotiation process and came up with five guiding principles: justice, sustainability, efficiency, participation and accountability. In this context, critiques of modernisation and its neglect of broader social and environmental concerns became an integral part of the debate. It was hoped that with the ground-breaking work of the WCD in 2000, the controversies of the past were buried, which they were for a period, symbolised by the drop in funding by the World Bank (Baghel and Nüsser 2010; Moore et al. 2010).

However, with the spectre of global warming, dam building has once again become en vogue as a climate-friendly technology, validated by global concerns for the reduction in carbon emissions. In the course of this latest disjuncture in the debate, large dams are now seen in a different light as mechanisms for mitigating against the adverse effects of climate change through their low emissions during power production ('blue becomes green'). Not only has climate change brought a new assertiveness to the benefits of hydro-dams, they have also given birth to the emergence of climate economies, whereby funding is now relatively more easily available in the 'free market' through trade mechanisms of carbon emission certificates. This trend becomes apparent in huge financial support whereby the international carbon trading scheme Clean Development Mechanism (CDM) subsidises dam building in China and India as 'carbon offsetting dams' (Erlewein and Nüsser 2011, p. 293). In this context, the ideals of neoliberalism have taken hold, with the private sector becoming a key player in the funding and construction of large dams.

As even this brief overview of the large dam debate suggests, there are challenging questions, but no easy answers on such a complex topic. The contributions in this volume accordingly reflect the debate and address it from a multiplicity of perspectives. The varied geographical case studies, historical, social and economic settings within which the authors present their individual contributions, shed light on specifics and highlight the deep interconnections and disjunctures between different aspects of the large dams debate.

1.4 The Contributions in This Volume

The contributions in this book explore the multidimensional asymmetries of scale, time and directions in the Asian large dam controversy. Using a Geography of Science approach, Ravi Baghel outlines a genealogy of the hydraulic mission and the evolution of the meaning of dams in the South Asian, especially the Indian context (Chap. 2). He argues that large dams are an anatopism in the countries of the Global South. He uses the example of the Tennessee Valley Authority in the USA and the problems of its transfer to the Indian context. Miriam Seeger refers to another important facet of technological hydroscapes, namely, the large-scale diversion of water across watersheds (Chap. 3). Using the example of the South-north Water Diversion, Seeger concludes that the massive transformation of drainage patterns does not only serve to provide a regular water transfer to the economic centres of China, but that they also serve as an instrument to strengthen China's international position. Rohan D'Souza demystifies the view that large dams are purely expert driven techno-economic artefacts which are beyond politics (Chap. 4). In contrast he argues that large dams are deeply embedded in a political context and that their outcomes are often directed towards realising specific political goals.

Based on a case study of the flood-prone Damodar River, a tributary of the Ganga, Kumkum Bhattacharyya and Michael J. Wiley depart from the environmental dimension of the transformation of fluvial environments (Chap. 5). They show that despite the construction of four multipurpose dams to reduce flood peaks, the lower Damodar Valley is at risk of relatively infrequent but more devastating floods. Bangladeshi refugees as new riparian communities are especially threatened by these changes in the fluvial environment. Nirmalya Choudhury provides an analysis of Environmental Impact Assessment follow-up in India (Chap. 6). He argues that the consistent decline in the use and application of this monitoring-evaluation-management cycle stems from the instrumental role of environmental clearance in India. This also includes the market-friendly macro-political environment and the prevalence of weak regulatory institutions. Departing from the perceived dichotomy of tradition and modernity, Hanna Werner discusses the complexity of the large dams controversy in India's developmental politics and draws particular attention to the case study of the Tehri Dam (Chap. 7). In the final chapter, Alexander Erlewein analyses the latest turn in the large dams debate in the context of global climate change (Chap. 8). Like a phoenix from the ashes, hydropower dams are now re-evaluated in terms of their potential to reduce greenhouse gas emissions. New institutions such as the Clean Development Mechanism (CDM), one of the flexible measures of the Kyoto Protocol, have fuelled the emergence of new 'climate economies'. As a result, many new carbon-offsetting hydropower dams have spread in the countries of the Global South. Taking into account the prevailing asymmetries of causation and burden sharing in climate change, Erlewein discusses whether this boom should be interpreted as 'carbon colonialism'.

Acknowledgement The author would like to express his gratitude to Paul Roden (Heidelberg) for careful proof-reading of the manuscript.

References

Adger WN, Benjaminsen TA, Brown K, Svarstad H (2001) Advancing a political ecology of global environmental discourses. Dev Change 32(4):681–715

Agrawal A (2005) Environmentality: technology of governments and the making of subjects. Duke University Press, Durham/London

Appadurai A (1996) Modernity at large. Cultural dimensions of globalization. University of Minnesota Press, Minneapolis

Baghel R (2013) Knowledge, power and the transformation of rivers: the spatial, governmental and subjective dimensions of expert discourse on river control in India. PhD thesis, Universität Heidelberg

Baghel R, Nüsser M (2010) Discussing large dams in Asia after the world commission on dams: is a political ecology approach the way forward? Water Altern 3(2):231–248

Baruah S (2012) Whose river is it anyway? Political economy of hydropower in the eastern Himalayas. Econ Pol Wkly 47(29):41–52

Blackbourn D (2006) The conquest of nature: water, landscape and the making of modern Germany. W.W. Norton, New York/London

Brewer N (2007) Made by China: damming the world's rivers. World Rivers Rev 22(3):8–9

Bryant RL, Bailey S (1997) Third world political ecology. Routledge, London/New York

Dai Q (1994) Yangtze! Yangtze! Earthscan, London/Toronto

Dai Q (1998) The three gorges project: a symbol of uncontrolled development in the late twentieth century. In: Dai Q et al (eds) The river dragon has come! the three gorges dam and the fate of China's Yangtze river and its people. M.E. Sharpe, London/New York, pp 3–17

Dittrich C (2004) Widerstand gegen das Narmada-Staudammprojekt in Indien. Geographische Rundschau 56(12):10–15

Drèze J, Samson M, Singh S (eds) (1997) The dam and the nation. Displacement and resettlement in the Narmada valley. Oxford University Press, New Delhi

D'Souza R (2006) Drowned and dammed: colonial capitalism and flood control in eastern India. Oxford University Press, New Delhi

Erlewein A (2012) Energie aus dem Himalaya: Ursachen und Folgen des aktuellen Wasserkraftbooms in Himachal Pradesh, Indien. Geographische Rundschau 64(4):26–33

Erlewein A, Nüsser M (2011) Offsetting greenhouse gas emissions in the Himalaya? Clean development dams in Himachal Pradesh, India. Mt Res Dev 31(4):293–304

Florini A (2000) The third force: the rise of transnational civil society. Carnegie Endowment, Washington, DC

Gadgil M, Guha R (1994) Ecological conflicts and the environmental movement in India. Dev Change 25(1):101–136

Gleick PH (1998) The status of large dams: the end of an era? In: Gleick PH (ed) The world's water 1998–1999. The biannual report on freshwater resources. Island Press, Washington, DC, pp 69–104
Gleick PH (2012) China dams. In: Gleick PH (ed) The world's water. Vol. 7. The biannual report on freshwater resources. Island Press, Washington, DC, pp 127–142
Iyer RR (2003) Water - perspectives, issues, concerns. Sage, New Delhi/Thousand Oaks
Iyer RR (2007) Towards water wisdom: limits, justice, harmony. Sage, New Delhi/Thousand Oaks
Kaika M (2006) Dams as symbols of modernization: the urbanization of nature between geographical imagination and materiality. Ann Assoc Am Geogr 96(2):276–301
Kaisti H, Käkönen M (2012) Actors, interests and forces shaping the energyscape of the Mekong region. Forum Dev Stud 39(2):147–158
Khagram S (2003) Neither temples nor tombs: a global analyses of large dams. Environment 45(4):28–37
Khagram S (2004) Dams and development: transnational struggles for water and power. Cornell University Press, Ithaca/London
Kreutzmann H (2004) Staudammprojekte in der Entwicklungspraxis: Kontroversen und Konsensfindung. Geographische Rundschau 56(12):4–9
Li H, Waley P, Rees P (2001) Reservoir resettlement in China: past experience and the Three Gorges Dam. Geogr J 167(3):195–212
McCully P (2001) Silenced rivers: the ecology and politics of large dams. (enlarged and updated edition). Zed Books, London/New York
Ministry of Water Resources (2012) Brief note on organizational history of ministry of water resources. New Delhi, http://wrmin.nic.in/index2.asp?slid=283&sublinkid=542&langid=1
Molle F (2009) River-basin planning and management: the social life of a concept. Geoforum 40(3):484–494
Molle F, Mollinga PP, Wester P (2009) Hydraulic bureaucracies and the hydraulic mission: flows of water, flows of power. Water Altern 2(3):328–349
Moore D, Dore J, Gyawali D (2010) The world commission on dams + 10: revisiting the large dams controversy. Water Altern 3(2):3–13
Nilsen AG (2010) Dispossession and resistance in India: the river and the rage. Advances in South Asian Studies 16. Routledge, London/New York
Nüsser M (2003) Political ecology of large dams: a critical review. Petermanns Geogr Mitt 147(1):20–27
Nüsser M, Schmidt S, Dame J (2012) Irrigation and development in the Upper Indus Basin: characteristics and recent changes of a socio-hydrological system in Central Ladakh, India. Mt Res Dev 32(1):51–61
Roy A (2001) The greater common good. In: Roy A (ed) The algebra of infinite justice. Penguin, New Delhi, pp 43–141
Swyngedouw E (1999) Modernity and hybridity: nature, regeneracionismo, and the production of the Spanish waterscape, 1890–1930. Ann Assoc Am Geogr 89(3):443–465
Swyngedouw E (2009) The political economy and political ecology of the hydro-social cycle. J Contemp Water Res Educ 142(1):56–60
Viviroli D, Weingartner R (2008) 'Water towers' – a global view of the hydrological importance of mountains. In: Wiegandt E (ed) Mountains: sources of water, sources of knowledge. Springer, Dordrecht, pp 15–20
Viviroli D, Dürr HP, Messerli B, Meybeck M, Weingartner R (2007) Mountains of the world, water towers for humanity: typology, mapping, and global significance. Water Resour Res 43:1–13
Watts M, Peet R (2004) Liberating political ecology. In: Peet R, Watts M (eds) Liberation Ecologies, 2nd edn, Environment, development, social movements. Routledge, London, pp 3–47
Wittfogel K (1957) Oriental despotism: a comparative study of total power. Yale University Press, New Haven
World Commission on Dams (2000) Dams and development: a new framework for decision-making. Earthscan, London
Yardley J (2007) Chinese dam projects criticized for their human costs. New York Times, November 19, 2007

Chapter 2
Misplaced Knowledge: Large Dams as an Anatopism in South Asia

Ravi Baghel

Abstract Even though South Asia accounts for a large proportion of the dams in the world, they have seldom been examined at the subcontinental scale, with most scrutiny confined to specific projects. Large dams are not merely functional technologies but come invested with a broad range of meanings. Using a Geography of Science approach, this chapter attempts to create a genealogy of the evolution of the meaning of dams and identify the ways in which they have been influenced by the spaces in which this technology developed. Going beyond a simplistic local-global opposition, I argue that large dams are technological attempts to recreate the landscape in the image of other idealised spaces. In many ways, this recreation is fundamentally at odds with local conditions and makes large dams an anatopism in South Asia.

Keywords Large dams • Tropicality • Hydraulic mission • Genealogy • South Asia

2.1 Introduction

South Asia accounts for a substantial proportion of the large dams identified by the World Commission on Dams (WCD 2000), with recent figures showing that India alone has 4,711 large dams with another 390 under construction (CWC 2009). Whereas the ecological effects of large dams have been studied frequently within the context of individual projects, or at the scale of river basins, their suitability at the subcontinental scale of South Asia has not received adequate scrutiny. This

---

R. Baghel, M.A. (✉)
Department of Geography, South Asia Institute, Heidelberg University,
Im Neuenheimer Feld 330, 69120 Heidelberg, Germany
e-mail: baghel@uni-heidelberg.de

M. Nüsser (ed.), *Large Dams in Asia*, Advances in Asian Human-Environmental Research, DOI 10.1007/978-94-007-2798-4_2,
© Springer Science+Business Media Dordrecht 2014

mismatch with the broader environmental setting has been explored at times by activists, as well as by some experts like Kapil Bhattacharjee (Nandy 2001). For the most part, however, critics of large dams have instead focused their attention on the socioecological consequences of *specific* projects. This meant that proponents of large dams have had no need to defend large dams as a technological or even rational choice, with the objections remaining confined to ethical and economic ones.

It is essential to analyse large dams not as mere technological artefacts but as massive infrastructure with social, historical and cultural dimensions, deeply implicated in human relations with the environment. This is necessary in order to come up with a much more general critique of our relation with this technology, rather than with a simple evaluation in terms of cost-benefit, or ecological impact. Secondly, broadly considered, large dams are not simply about generating electricity or aiding in irrigation but have many symbolic aspects in addition to their functional ones. When seen in terms of our relation with the environment, this technology is based on the intention to transform the landscape, to make it more productive, more suitable for human progress. This might appear to be self-evident and too obvious to be noted, but it is this very property, hiding in plain sight, that deserves critical examination. In assessing large dams in terms of human relations of meaning with the landscape, this chapter complements other contributions in this volume that emphasise the role of dams in our social and political relations and in examining the asymmetric relations between expert knowledge and local knowledge that become apparent through exercises aimed at promoting public participation in Environmental Impact Assessments (Choudhury, Chap. 6, this volume; D'Souza, Chap. 4, this volume).

Though the temporal concept of anachronism, something that belongs to a different time, is relatively familiar, its geographic counterpart is lesser known. Anatopism is the analogical idea of objects that are 'out of place', or appear to belong to another space. I argue that dams are such an anatopism in South Asia and are flawed attempts to recreate an ideal river whose conception is derived from elsewhere. This examination proceeds with the premise that the technology of large dams is neither objective nor universal and is inflected with the spatial and historical contexts through which it has evolved. By focusing on what large dams mean in spatial terms, as a form of human relation with the inhabited space, the assumptions about the ideal space and their contradictions with the experienced environment become apparent. This can help in moving beyond entrenched positions and move closer to the heart of the dams debate. An emphasis on the meaning attributed to large dams in social, cultural and historical terms can create the basis for a more creative engagement, as opposed to a merely reactive one. At this point, it is important to mention that it is not the development of large dams as a technology to correct or alter the flows of rivers that is being examined here, but rather how the problem was constructed to which large dams were seen as the solution.

This chapter follows the tradition of a critical examination of technology in terms of human relations that are encompassed under the broad rubric of Science and Technology Studies (STS). However, the emphasis here is on the spatial or geographical aspect, and focuses on the influence of space on the evolution of the technology of large dams and subsequently its impact on the space to which this technology is applied. A Geography of Science (GoS) approach is the conceptual framework used to relate the spatial context to the technology of large dams. The GoS perspective is employed in this chapter partially as a critique of the technology of large dams. However, merely identifying this particular technology as co-produced by its geography (and therefore lacking 'objectivity') is not the intent of this chapter. My argument here proceeds in two parts, with the first part focusing on the framing of the problems to which large dams are offered as solutions. The framing of the general problem in terms of the so-called hydraulic mission (Molle et al. 2009) is placed in its historical context, with a focus on the interplay of knowledge and space legitimised by notions of environmental determinism. The second part examines their efficacy as solutions. In the case of South Asia, this chapter argues there are several specific environmental features that make this technology, especially in the form it is normally applied, inappropriate in most locations.

The next section acts as a preamble introducing the conceptual framework and points out the possibilities inherent in a spatial examination of large dams, as well as the hydraulic mission that drives their proponents. Following this, a few contradictions inherent within the hydraulic mission are highlighted. The third section suggests some practical objections that arise from the incompatibility of the hydraulic mission with the geographic context in which it is pursued. This chapter is concluded by a discussion of the possible alternatives to the hydraulic mission and the possibilities for a more creative critique of large dams.

2.2 Placing Technology

The examination of the social and cultural influences on technologies has a long history in which the work of Thomas Kuhn on the scientific method (Kuhn 1996) is seen as a milestone. Following this, the fields of Sociology of Scientific Knowledge (SSK) and the broad area of Science and Technology Studies (STS) have achieved prominence in critical studies of technology. While its social, cultural and economic origins have received considerable attention, a focus on the spatial production of knowledge is relatively recent. The most fruitful engagement with the spatial aspect of science and technology has emerged from the field of GoS.[1] This approach relies upon the premise that knowledge is produced in a particular place, and, *where*

[1] A note on terminology is useful here, as there is a multitude of possible usages that are current, the most prominent being "Geography of Science" and the derivative "geographies of science" and the broader term "geography of knowledge". This paper uses the nomenclature "Geography of Science" as a device to ensure consistency and ease of understanding, but it is intended to encompass the broader idea of "knowledge". The implication of plurality is inherent when this term is used so broadly, and therefore, I do not use "geographies" as an additional rhetorical device.

knowledge is produced, has an influence on *what* is produced. As Livingstone asserts:

> Scientific knowledge is made in a lot of different places. Does it matter where? Can the location of scientific endeavor make any difference to the conduct of science? And even more important, can it affect the content of science? In my view the answer to these questions is yes (Livingstone 2003, p. 1).

This emphasis on geography has provided this field with a variety of key concerns, each with varying levels of overlap with other fields like Science and Technology Studies (STS) and Sociology of Knowledge. These fields assert that knowledge cannot be value neutral, and that the sociocultural milieu in which the knowledge is produced affects both the form and content of knowledge. Examples of this can be seen for instance in the role of social status, not just of the scientist, but especially of the audience, in establishing the validity of scientific knowledge (Shapin and Schaffer 1989). What distinguishes GoS is the emphasis on the role of space and place in the production of scientific knowledge.

Researchers working on GoS have confidently asserted that all scientific knowledge originates in a place and is influenced by it (Livingstone 1995, 2003). This is in itself a questioning of science, and by inference of technology, as what Nagel (1986) has called the 'view from nowhere' (cited in Shapin 1998, p. 5). Though a GoS perspective is inherently critical through its questioning of the universalist claims of the technology of large dams, the influence of geography on this technology is not remarkable in itself. What is remarkable is that the universalist claims of this technology have been so readily accepted. Once the spatial influences on the form and content of this technology are acknowledged, then its applicability to other locations can no longer be taken for granted and must instead be scrutinised afresh. Before proceeding to give the broad contours of the GoS, it is helpful to consider some examples of what it is *not*. An example of a trivial and simplistic way of dealing with the connections between place and science is that adopted by Dorn (1991). This work tries to identify ultimate origins of certain kinds of knowledge and uses a kind of geographical cataloguing of technologies and inventions, using an eclectic mix of cultural and political economic determinism, informed by a eurocentric perspective. This kind of self-indulgent cataloguing contributes little to our understanding and is often suspect in what is identified as important, as well as in its attribution of unique origins to technology. GoS on the contrary relies on a richer understanding of geography, viewing it as more than merely a set of fixed cartographic coordinates; instead, space is understood in terms of systems of relations between the environment, place and people.

Furthermore, GoS should not be confused with environmental determinism, or closely related arguments that focus on the geography, or more precisely spatial properties, as a sort of backcloth that creates the basis for what kind of knowledge is possible. An elaborate example of such recent work that has attained considerable popularity is Jared Diamond's *Guns, Germs and Steel* (1998), in which he attempts to explain away current inequalities among regions of the world as a product of environmental differences and geographic influences on societies. Blaut (1999) has provided an excellent and detailed critique of this (and similar) work by placing it in a tradition of eurocentric history whose theological and racial underpinnings have retreated into the background, leaving environmental determinism as the 'scientific' explanation of European superiority.

Sheppard identifies another more contemporary function of this type of work, as a defence or justification of globalising capitalism:

> Diamond's geographical imaginary... leans toward a teleological account of economic development in which 'geography' disrupts the otherwise flattening playing field of globalizing capitalism (Sheppard 2011, p. 47).

Sheila Jasanoff has used the idea of *co-production* in order to avoid privileging either sociocultural or techno-scientific determinism in the study of technology (Jasanoff 2004, p. 20). Extending this idea of co-production naturally focuses attention on the influence of the geographical, cultural and social context on the production of the technology. Thus, large dams cannot be understood as abstract technologies but need to be conceptualised as co-produced by their context. This suggests that whereas the ideas on their technical functioning have arisen out of situated practice and experience, the meanings with which dams are invested are very much produced through the social, political and historical context in which they evolved.

It is easy to use the idea of the influence of place on the evolution of large dams for a parochial defence of the local. Therefore, an additional caveat needs to be offered, which is that a critique based on GoS is subtly different from a simplistic account that places local and nonlocal knowledge in opposition, where one is seen as necessarily better and the other is denigrated. An attention to the geographical influences *contextualises* the technology of large dams, without being a polemic attack on this technology. Paying attention to the context and bringing the universality of this technology into question does not imply its being universally inappropriate; there certainly are settings in which such technology may be valuable. The next section attempts the tightrope walk necessary to separate an *attention* to place in the evolution of knowledge from the *reification* of place as a determinant of the value of knowledge.

2.3 A Genealogy of a Mission

Different kinds of dams have been built all over South Asia and are mentioned even in ancient sources like the *Rigveda*, while elaborate rules on their maintenance are prescribed in the administrative manual *Arthashastra* dating back more than two millennia. However, the revival of dams can be traced back to the British in colonial India, and the construction of large dams truly took off in 1947, following the formation of the modern independent nation-states of South Asia. Colonial projects were primarily oriented towards profit extraction and revenue generation, giving these efforts a specific character that has been called 'colonial hydrology' (D'Souza 2006). The succeeding period had several features that distinguish it from the colonial era, most remarkable of which is perhaps the sense of urgency with which dam building was pursued. This urgency was characteristic of this period and could be found in many countries, but especially in newly decolonising nation-states like India, Pakistan, Egypt and China. This overarching idea of transforming rivers and using up every drop of their water in the service of humanity has been called the 'hydraulic mission' (Molle et al. 2009).

The primary purpose of large dams has usually been identified as one or all of the three functions of flood control, irrigation and hydropower generation. In South Asia, the emphasis has traditionally been on irrigation, though it is now gradually shifting towards hydropower generation. All these efforts can be summed up as an attempt to create the ideal river which is predictable, useful and completely in the service of humanity. This ideal however is by no means the only one possible, since rivers have always been deeply implicated in terms of spiritual, aesthetic and cultural relations with nature, with a correspondingly differing conception of what the ideal river is or ought to be.

2.3.1 Theological Underpinnings

The origins of the normative ideal the 'hydraulic mission' attempts to recreate have been traced back to north-western Europe. Tuan Yifu (1968) traces the origins of the modern understanding of rivers to seventeenth-century Natural Philosophy. The key objective of this field of knowledge was to describe the world as the creation of a supreme being, considered omniscient, who had arranged a perfect nature for the providence of mankind. Science or Natural Philosophy at this time was not generally understood as an objective search for the truth but more as an endeavour that sought to explain the works of the creator. An assumption that was implicit in this project was that of God being a merciful provider for the needs of his chosen people. As these ideas were developing in north-western Europe, the local experience of rivers as even flowing, seasonal, predictable and perennial became the prism through which water became understood. Linton (2008, 2010) has extended these ideas to the development of the now universally recognised model of the hydrologic cycle and the founding of the International Hydrological Association. He argues that this period was the beginning of an era in which water became a 'modern abstraction' and was stripped of its social, cultural and spiritual properties. The ease with which dams and other modifications of rivers and water systems fit into this model is also remarkable, and its influence on the ideas guiding the iconic Tennessee Valley Authority (TVA) that came to life in this context are worth mentioning.

The physical geography of Europe, where modern hydrology was taking root, also left traces in all future conceptions of rivers and riverscapes, since defining the normal was done by taking as the standard the experience of water in parts of Europe. This, coupled with the theological underpinnings, led to a denigration not just of other experiences with water but also of other people living in places that had an association with water that did not comply with the European normative expectations (Linton 2010).

> The notion that 'all the land' was well supplied with water was not unique to Ray. Whether out of ignorance or out of enthusiasm, the geographical fact of aridity was conveniently overlooked by proponents of what might be called the 'sacred' hydrologic cycle (Linton 2008, p. 633).

Large dams become desirable only when aridity is seen as an aberration, and not a normal fact of life. In the light of its theological underpinnings, large scale engineering of rivers was a way of reproducing the Edenic ideal of abundant water, which depended on a disparagement of other experiences of water. These perspectives fit in neatly with the prevailing notion of the civilising mission, and a correction of both moral and material nature was seen not only as necessary but also as an obligation upon the colonisers.

2.3.2 *Tropical Nature*

David Arnold has used the term 'tropicality', in a conscious parallel to the concept of orientalism developed by Edward Said (1979), to describe the imagination of the landscape of the *Other* (Arnold 2000) that is seen in the process through which the tropics were perceived to be another space with their own character, distinct from other parts of the world. Of course, there was no monolithic way in which the tropics were imagined, but common to all such ideas was the sharp distinction made between the tropics and the rest, and the attribution of shared qualities to all tropical areas. These notions were by no means always negative, and Arnold (2004) argues that a negative view of India as a place of disease and death was produced through the descriptions and experiences of colonial travellers and was distinct from the earlier view of tropical nature as abundant, fertile and rich. He further asserts that it was this changed perception of India as a ruined and desolate 'deathscape' that became the justification for the various administrative projects of 'improvement'. These early attempts at improving upon a natural environment perceived to be perverse were simply the precursors of what later became the project of modernisation.

A chance combination of factors, both human and natural, may have further contributed to the changed perception of India from a prosperous land of riches to an impoverished disaster zone. Mike Davis connects the especially severe famines of the late nineteenth century to droughts caused by a very strong El Niño/La Niña-Southern Oscillation (ENSO) (Davis 2000). This extreme weather event was compounded by the breakdown of traditional administrative structures of famine prevention, the *increasing* exports of food during this period and the incorporation into world commodity chains. The change of land use from subsistence farming to monoculture of cash crops across a large area also added to the vulnerability of local populations. Davis has called this process the 'modernisation of poverty'. The simplistic understanding that famines are caused by a lack of food, or by agricultural failure, has been challenged successfully (Sen 1981). However, the perception of Indians being at the mercy of nature has endured.[2]

Further, the effects of tropicality were not just limited to a representation of the tropical environment but also extended to descriptions of the *moral* nature of its inhabitants. Livingstone commented on the briefly flourishing pseudoscience of ethnoclimatology and described a 'moral climatology' that was used to supply a scientific justification to the presumed connections between race, space and moral character (Livingstone 1991, 2000, 2002). The assumed connection between moral qualities like fatalism, superstition, passivity and the tropics was effective in creating the rationale for the governance of the colonised. These ideas not only served to legitimise the 'civilising mission' but also supported a depiction of locals as lacking agency and volition. These views were quite tenacious, and as late as 1965, the award-winning work of Nirad Chaudhuri (1965) firmly ascribed many perceived failings of India and Indians like laziness, passivity and ignorance to its tropical location. In combination with the natural disasters that were observable, this contributed to a view of the necessity of state interventions to ensure food self-sufficiency. This is a problem that was considered especially acute until the onset of the green revolution, whose ostensible success is often cited as justification for large dams (Dharmadhikary 2005).

With the beginning of decolonisation, it seems natural that the transformation of both moral and material nature went hand in hand. And the antidote to the colonial experience was seen as the active transformation of the environment, in this case rivers, in order to achieve a reversal in fortune. These ideas driving dam building are perhaps brought out most eloquently in Nehru's famous proclamation of dams as the 'new temples of modern India'. Far more telling than the use of this phrase is the resonance it found among government officials at large and also the broader public. The connection between religion, superstition and resultant 'backwardness' is a recurring one in the thought of Nehru (2004). Against this backdrop, the move away from sacred rivers, towards dams as temples, in order to create a 'modern' India was perhaps inevitable.

2.3.3 Environmental Determinism

One of the first acts of the parliament of independent India was to create a constitutional act to create the Damodar Valley Commission modelled on the TVA, which was very intimately connected to its myth of democratic and socially

[2]A billboard from the late 1990s at the site of a dam project, under the headline "Sardar Sarovar Project: A gift to the nation", states: "India must be self-sufficient in food to avoid going to the world with a begging bowl. Country needs many more dams to harness waters going waste to sea after disasterous damage of flood" (sic).

transformative development (Klingensmith 2007). This urgency to transform the landscape has many influences, but the shift in focus from dams that were profitable undertakings, to dams that were almost magical in their emancipatory potential, is remarkable. An engineer working on the Bhakra project described how the arrival of the Americans at the Bhakra project, initiated by the British, transformed the morale and the motivation of Indians, ushering in a patriotic pride in a project that was ostensibly going to transform the nation (Singh 1998). What connects this crusade to transform the landscape to fulfil the potential of the nation to the modern South Asian conception of rivers is the idea of environmental determinism.

In the urgency of decolonisation, all nations of South Asia joined the race for modernisation and tried to overcome the factors that they considered responsible for their poverty and backwardness. While racial explanations were no longer tenable, environmental determinism and cultural failings were the most acceptable framings of perceived backwardness. Whereas environmental determinism can be summarised as the argument that geography is destiny, dams were an attempt to modify the landscape to overcome destiny by correcting perceived shortcomings of the environment. In this context, those who advocated adaptation, or less drastic solutions to the problems of river control, were dismissed as backward and fatalistic. Where such allegations did not work against well-known critics of dams, they were labelled anti-national.[3]

Any genealogy is only a narrative imposed on what were essentially disparate events; however, the evolutionary path described here is an attempt to account for the fervour with which dam building was pursued. The problematic origins of this zeal might have been excusable, if they had not had the terrible consequences they did when applied to the South Asian context. The next section suggests some of the major contradictions that resulted from these attempts to transform the environment to create the imagined ideal rivers.

2.4 Hydraulic Mission and Its Contradictions

The examination of large dams conducted here does not focus on their technology, but on their logic. Due to variability of design and diverse practices, the performance of large dams is naturally equally varied. Further, the construction of all large dams does not necessarily derive from the hydraulic mission described earlier, and there is a long history of indigenous dam building by local kings and feudal lords with varied outcomes (Morrison 2010). However, the hydraulic mission as a national

[3]This practice has endured to the present day. In the 1950s Kapil Bhattacharjee was accused of being a Pakistani spy (Nandy 2001). In more recent times, many renowned activists like Medha Patkar and Arundhati Roy have been called anti-national and alleged to have connections to unnamed foreign powers. This reaction is all the more interesting, because of the unquestioned identification of dams with the nation and of meanings given to dams that extend far beyond their function.

project, driven by the imperatives described earlier, results in the set of common contradictions listed below, due to its very logic. Though there are many problems with the hydraulic mission, some become especially prominent when a spatial analysis within an STS framework is pursued.

2.4.1 *Placeless Origins*

Many of the key principles that underlie the thinking behind large dams rely upon the illusion of the placelessness of science that has been challenged in recent times (Livingstone 2003; Withers 1995; Naylor 2005). Many of the hydraulic principles upon which the hydraulic mission relies, like the hydrologic cycle (Tuan 1968; Linton 2010) and the river basin (Molle 2009), developed in north-western Europe and continue to bear traces of this geographical and intellectual context. One of the effects of this history is the reliance upon even flowing perennial rivers as the normative ideal, one of the consequences of which has been the denigration, or dismissal, of aridity and seasonality of other experiences of river regimes. Mike Davis (1995) argues that the 'humid fallacy' was a culturally specific prejudice based on the experience of annually and seasonally predictable precipitation in parts of Europe and the east coast of the USA. This norm of annual averages was extended to the very different Mediterranean littoral environment systems of Southern California, Chile, Australia and the South African Cape, which are characterised by extreme events and have oscillating dry and wet periods on the scale of 7–12 years. Calder (1999) has made similar arguments about the incompatibility of the normative ideals of the hydraulic mission with the hydrological experience in parts of Africa.

The lack of attention to the origins and the associated presumption of uniformity of experience directly contradict the diverse systems that exist in practice. Perhaps if the specificity of all these diverse hydrologic experiences had been recognised, then the misadventures in 'correcting' the water balance would not have had such disastrous environmental consequences in many parts of the world. Unfortunately, even today, hydrologic models continue to rely on concepts like annual stream flow and average rainfall, concepts which have little meaning in places where the 'average' is the exception rather than the norm.

2.4.2 *Universal Replicability*

In addition to the perceived placelessness of origins, another related assumption implicit in the hydraulic mission is the universal replicability of this technology. As the leading proponent of the TVA model, its chairman David Lilienthal proclaimed:

> I write of the Tennessee Valley, but all this could have happened in almost any of a thousand other valleys where rivers run from the hills to the sea… In China and in India there are

> just such rivers... rivers that in the violence of flood menace the land and the people, then sulk in idleness and drought – rivers all over the world waiting to be controlled by men – the Yangtze, the Ganges, the Ob, the Parana, the Amazon, the Nile (Lilienthal 1944, pp. 1–2).

This has meant that not only have even flowing, perennial rivers been unquestioningly accepted as the normative ideal for South Asian rivers; the means for achieving this flawed ideal have also remained similarly unquestioned. This fallacy is especially relevant when it is considered with respect to some conditions that are characteristic to the subcontinent. For instance, due to the unpredictability and high intensity of rainfall, the operation of reservoirs becomes difficult. Reservoirs may have to be emptied for the safety of dams, flooding downstream communities.

Many South Asian rivers, especially those originating in the Himalayas, have a very high sediment load. The resulting siltation of large dams has affected many reservoirs reducing their functionality. This is apart from the much greater ecological costs of keeping back as much as 25 % of sediments trapped within the reservoirs (Vörösmarty et al. 2003) that deprives downstream organisms and communities of nutrients and soil. However, this has for the most part been blamed on poor land management practices upstream, which has meant continued support for large dams, and indeed often the construction of additional check dams designed for the express purpose of reducing siltation. Not only does siltation lower the performance of dams, but this also has consequences downstream. A large study of rivers in India found 'dramatic reductions in sediment load in the tropical river basins... beyond the fold of assignable natural variability' (Panda et al. 2011, p. 108). The largest effect was found for the Narmada River, which the authors ascribe to dams. Further, they argue that the trapping of sediments has also significantly contributed to coastal erosion.

An additional factor especially relevant for South Asia is the high population density in many parts. This drastically increases the number of people displaced due to flooding by reservoirs, as compared to other parts of the world. Surprisingly enough, there are no accurate figures on the number of people thus affected, but their number has been estimated to be anywhere between 10 and 30 million, in India alone. Even when dams are built in areas with low population density, as in the Indian state of Arunachal Pradesh, those affected tend to be smaller tribal communities, significantly exacerbating the impact on social cohesion and cultural practices.

2.4.3 Ahistorical Development

Technologies are usually seen as being value neutral, and often the position of technologies, being neither good nor bad, masks the actual ideological traces within them. Thus, large dams appear to emerge from nowhere, and the connection to the context, in which the TVA emerged, is subsumed under its 'myth' (Hargrove 1994; Klingensmith 2007).

These same ideas when coupled with industrial capitalism, as in the case of the American south, then not only led to dams as a solution but also to the reliance upon 'largeness' as an additional desirable quality. Ironically the advent of large dams in the USA happened during one of the periodic crises of capitalism in the depression era, which manifested in a state-financed development of the TVA rather than one financed by private capital, as had previously often been the case for large dams in the USA. This reliance on state driven and large projects has continued to the present day, accompanied by an unquestioned belief in state directed improvement of life through drastic interventions.

2.4.4 Nationalised Environments

Instead of recognising the highly diverse ecological conditions across the entire South Asian region, governments in the region have seen nature as undifferentiated and in national terms (Baghel and Nüsser 2010, p. 241). This is evident from statements that speak of India as a water surplus country and in similarly deluded proposals to transfer water from 'water surplus basins' to 'water-deficient basins' through the interlinking of rivers (Iyer 2007, p. 50). Indeed, while the Indian proposals seem to have once again receded from the public imagination and from the policy planning process, in China two massive water diversion projects based on similar faulty premises are already fairly well advanced and beyond the planning stage (see Seeger, Chap. 3, this volume).

It has been suggested that this way of seeing the entire national territory as undifferentiated, or at least one in which variety is seen as an aberration, is directly related to the prerequisites of governance by the modern state (Scott 1999, 2006). From this, it can be argued that the hydraulic mission and the fallacy of a nationally homogeneous environment are essentially inseparable. The contradictions arising out of this lack of attention to ecological variability within the national territory are described later. A corollary of perceiving rivers as a national resource is that even their administration is done at the national scale, with the size of river control schemes being correspondingly large. Additionally, river planning at this scale contributes to conflicts over water sharing on both the international and interstate levels.

However, planning in terms of a single national environment is a direct contravention of the immense internal ecological variability of South Asia. In their classic work on the traditional water systems of India, Agarwal and Narain (1997) have identified at least 14 distinct ecological zones, each with a water management system that has evolved to suit local conditions. These diverse adaptations to local conditions include examples such as the *Zings* of Ladakh which are canal networks that make use of glacial melt, the *Kunds* of Rajasthan which are designed to store sudden and rare precipitation for long-term use and tank irrigation in parts of Southern India. In contrast with this, large dams are a single solution imposed upon a variety of requirements and local conditions. Although dams are in general engineered differently to suit differing structural requirements, the identical strategy of a reservoir containing impounded water, distributed by canals, leads to high water loss due to evaporation and salinity in arid areas (Sakthivadivel et al. 1999). In case of rivers that have high sediment loads, for instance, in parts of eastern India, this same strategy leads to rapid siltation and loss of soil fertility downstream.

This internal ecological variability is in direct contradiction to the administrative prerequisite of a 'nationalised environment' as discussed earlier. The resulting governmental need for an easily replicable national strategy, combined with an inability to recognise local variations in environmental conditions, leads to poor performance and high environmental costs. Smaller scale solutions adapted to local conditions on the other hand may have the potential for better performance.

2.5 Dams as Anatopism

Apart from the numerous contradictions mentioned above, South Asia is also very densely populated, which makes the creation of reservoirs that flood large areas of inhabited space especially disruptive. This has contributed to social and political conflict, as well as created millions of internal refugees. South Asia also has a long history of agriculture, and local practices of water management have necessarily become well adapted to regional conditions. However, a modification of river systems at such a large scale has had cascading effects on agriculture downstream, and many of these formerly successful adaptations have now become unsustainable. In case of some parts of South Asia, flooding has worsened due to poor drainage, unplanned releases of water for dam safety, etc. When combined with the high population density, the transformation of flood and river regimes has magnified the effects of extreme weather events to catastrophic extents.

These factors mean that the use of large dams as a technology is questionable in most areas, and in fact the use of a variety of technologies tailored to the region in general and to specific ecoregions in particular is essential. By giving due attention to the spatial nature of knowledge, as well as its embodiment, the governance of river systems in particular and of all natural cycles in general can be significantly improved. It is also essential to not only level critique at individual instances of large dams, in which case criticism is easily deflected by pointing to specific problems (and solutions). By looking not only at the local scale and at the global scale as, for instance, in the case of the WCD (2000), but also introducing a scale sensitive to diverse ecological variations, the case of large dams can be addressed much better (Baghel and Nüsser 2010).

This addresses the weakness of studies that analyse large dams, but at the national scale, because this homogenises the nation as one particular kind of ecological space and loses sight of the appropriateness of the technology to specific ecoregions. The fallacy of 'national environments' is instead one that perpetuates the construction of large dams, and by using this level of analysis, a very important aspect of ecological variability and thus the varying appropriateness of large dams to specific ecoregions is disregarded. Further, it is important not only to recognise space as having agency and variability but also to recognise the aspect of place which is dynamic on smaller time scales as compared to, e.g. geological changes in the biophysical space. As place includes not only the biogeophysical environment but also the notion of affect or emotional response to this space, it is important to recognise that not only do emotional responses towards the biophysical environment constantly change, but also they constantly acquire new meanings. This means, for example, that a river may have been seen as sacred earlier, and then as an industrial resource, and at another time perhaps as a source of aesthetic pleasure. A dam constructed on this very river would therefore not be constructed on the very same 'place', which means that even when a technology has been considered appropriate at one time, it may have become inappropriate due to a change in values and the dynamic nature of place.

This becomes readily apparent when the case of the demolition of many large dams across North America is taken into account. Dams that were once seen as a necessary evil have come to be seen as merely evil. This has often been a result of changing responses to the river from an industrial source to an aesthetic one, the idea of a place as being part of a salmon run, revaluation of species (e.g. the process of the former 'malarial swamps' becoming reframed as 'wetlands with biodiversity'). The changing character of place and the variability of space therefore are both essential factors to be considered when analysing the appropriateness of large dams.

Due to the legacy of a normative ideal of rivers that has its basis in colonial era moral and theological judgement of spaces and its inhabitants, dams continue to be attempts at creating a utopia. However, this utopian mission is subtly inflected with an attempt to recreate the ideal spaces of colonisers in an attempt to overturn their judgements. The attempt to impose these spaces and spatial qualities from elsewhere have resulted in the incongruities of rivers that are out of place and an anatopism in the spaces where they have been recreated.

2.6 Conclusion

This chapter has discussed the phenomenon of large dams in terms of their fit as appropriate technologies with respect to the diverse environmental settings of South Asia. While it is difficult to do such a complex topic justice in the limited space of this chapter, this critique brings into focus a very important but overlooked aspect of large dams as a technology, namely, their connection with space and time. This technology needs to be put into its spatial and temporal context. A case has also been made in this chapter for acknowledging the dangers of ignoring the connections between space and knowledge. Geographers are especially well suited for addressing this gap, and bringing in a spatial perspective can enrich the dam debate while making newer approaches possible. These may also break the stalemate and enable shifts away from entrenched positions that are based on a simplistic understanding of dams.

By undermining the certainties of the proponents of the hydraulic mission, this chapter identifies the question 'Are large dams good or bad?' regardless of the answer, as fundamentally flawed for its disregard of both space and time. It is argued that the question instead needs to be changed to: 'Are large dams appropriate for this particular space and time?' This question addresses both the appropriateness of a large dam for a particular location, in terms of the dynamic constraints it imposes upon the functioning of a dam. In addition, this question also incorporates the equally important element of the appropriateness of this technology to a particular time, with respect to the changing values of a society. This recognises that the choice of such a massive and disruptive technological solution is never value neutral, and as a consequence, both proponents and opponents need to make the values that guide them explicit and ready for examination. The recognition of large dams as situated within a particular time and place further leads us to rethink human-environmental interactions from ideas of 'man and nature', 'man against nature'[4] to one of 'humans in nature' (Meyer-Abich 1990, 1996; Berkes and Folke 1998; Berkes et al. 2003).

Fervent proponents of dams have at times placed them in the context of Hammurabi's famous inscription where he claimed 'I have transformed the desert plains into fertile fields, given their residents fertility and abundance, and I have made the country an abode of delight' (Costa 1911 cited in Molle et al. 2009, p. 330). However, a proclamation symbolising the hubris of dam building with a more appropriate irony might instead be found in the words of Shelley:

> My name is Ozymandias, king of kings:
> Look on my works, ye Mighty, and despair! (Shelley 1826, p. 100)

Acknowledgement I would like to thank the Cluster of Excellence: Asia and Europe in a Global Context and the University of Heidelberg for the generous financial and intellectual support provided during the writing of this chapter. I must also thank Prof Nüsser for his mentoring and advice, Thomas Lennartz for contributing his valuable organisational skills and all my colleagues in the Department of Geography, South Asia Institute, for their encouragement and support.

References

Agarwal A, Narain S (1997) Dying wisdom: rise, fall and potential of India's traditional water harvesting systems. CSE, New Delhi

Arnold D (2000) "Illusory riches": representations of the tropical world, 1840–1950. Singap J Trop Geogr 21(1):6–18

Arnold D (2004) Deathscapes: India in an age of romanticism and empire, 1800–1856. Ninet Century Contexts 26:339–353

Baghel R, Nüsser M (2010) Discussing large dams in Asia after the World Commission on dams: is a political ecology approach the way forward? Water Altern 3(2):231–248

Berkes F, Folke C (eds) (1998) Linking social and ecological systems: management practices and social mechanisms for building resilience. Cambridge University Press, Cambridge

[4]The original sexist language is retained here to reflect the phallogocentric nature of such positions.

Berkes F, Colding J, Folke C (2003) Navigating social-ecological systems: building resilience for complexity and change. Cambridge University Press, Cambridge
Blaut JM (1999) Environmentalism and eurocentrism. Geogr Rev 89(3):391–408
Calder IR (1999) The blue revolution. Land use and integrated water resources management. Earthscan, London
Chaudhuri NC (1965) The continent of Circe: an essay on the people of India. Chatto & Windus, London
CWC (Central Water Commission) (2009) National register of large dams – 2009. Technical report. Government of India, New Delhi
D'Souza R (2006) Water in British India: the making of a 'colonial hydrology'. Hist Compass 4(4):621–628
Davis M (1995) Los Angeles after the storm: the dialectic of ordinary disaster. Antipode 27(3): 221–241
Davis M (2000) The origin of the third world. Antipode 32(1):48–89
Dharmadhikary S (2005) Unravelling Bhakra: assessing the temple of resurgent India. Manthan Adhyayan Kendra, Badwani
Diamond J (1998) Guns, germs, and steel: the fates of human societies. Norton, New York
Dorn H (1991) The geography of science. Johns Hopkins University Press, Baltimore/London
Hargrove E (1994) Prisoners of myth: the leadership of the Tennessee Valley Authority, 1933–1990. Princeton University Press, Princeton
Iyer RR (2007) Towards water wisdom: limits, justice, harmony. Sage Publications Inc., New Delhi/Thousand Oaks/London/Singapore
Jasanoff S (2004) Ordering knowledge, ordering society. In: Jasanoff S (ed) States of knowledge: the co-production of science and social order. Routledge, London/New York, pp 13–45
Klingensmith D (2007) "One valley and a thousand": dams, nationalism and development. Oxford University Press, New Delhi
Kuhn TS (1996) The structure of scientific revolutions, 3rd edn. The University of Chicago press, Chicago/London
Lilienthal DE (1944) TVA, democracy on the march. Overseas Editions, New York
Linton J (2008) Is the hydrologic cycle sustainable? A historical-geographical critique of a modern concept. Ann Assoc Am Geogr 98(3):630–649
Linton J (2010) What is water? The history of a modern abstraction. University of British Columbia Press, Vancouver
Livingstone DN (1991) The moral discourse of climate: historical considerations on race, place and virtue. J Hist Geogr 17(4):413–434
Livingstone DN (1995) The spaces of knowledge: contributions towards a historical geography of science. Environ Plann D Soc Space 13(1):5
Livingstone DN (2000) Tropical hermeneutics: fragments for a historical narrative: an afterword. Singapore J Trop Geogr 21(1):76–91
Livingstone DN (2002) Race, space and moral climatology: notes toward a genealogy. J Hist Geogr 28(2):159–180
Livingstone DN (2003) Putting science in its place: geographies of scientific knowledge. The University Of Chicago Press, Chicago/London
Meyer-Abich KM (1990) Aufstand für die Natur: Von der Umwelt zur Mitwelt, 2nd edn. Carl Hanser, München
Meyer-Abich KM (1996) Humans in nature: toward a physiocentric philosophy. Daedalus 125(3):213–234
Molle F (2009) River-basin planning and management: the social life of a concept. Geoforum 40(3):484–494
Molle F, Mollinga PP, Wester P (2009) Hydraulic bureaucracies and the hydraulic mission: flows of water, flows of power. Water Altern 2(3):328–349
Morrison KD (2010) Dharmic projects, imperial reservoirs, and new temples of India: an historical perspective on dams in India. Conserv Soc 8(3):182–195
Nagel T (1986) The view from nowhere. Oxford University Press, Oxford

Nandy A (2001) Dams and dissent: India's first modern environmental activist and his critique of the DVC project. Futures 33(8–9):709–731

Naylor S (2005) Introduction: historical geographies of science – places, contexts, cartographies. Br J Hist Sci 38(1):1–12

Nehru J (2004) The discovery of India. Penguin Books, New Delhi

Panda DK, Kumar A, Mohanty S (2011) Recent trends in sediment load of the tropical (Peninsular) river basins of India. Global Planet Change 75(3–4):108–118

Said E (1979) Orientalism. Vintage Books, New York

Sakthivadivel R, Thiruvengadachari S, Amerasinghe U, Bastiaanssen WGM, Molden D (1999) Performance evaluation of the Bhakra irrigation system, India, using remote sensing and GIS techniques (Research Report 28). International Water Management Institute, Colombo

Scott JC (1999) Seeing like a state: how certain schemes to improve the human condition have failed. Yale University Press, New Haven/London

Scott JC (2006) High modernist social engineering: the case of the Tennessee Valley Authority. In: Rudolph LI, Jacobsen JK (eds) Experiencing the state. Oxford University Press, New Delhi, pp 3–52

Sen A (1981) Ingredients of famine analysis: availability and entitlements. Q J Econ 96(3):433–464

Shapin S (1998) Placing the view from nowhere: historical and sociological problems in the location of science. Trans Inst Br Geogr 23(1):5–12

Shapin S, Schaffer S (1989) Leviathan and the air-pump: Hobbes, Boyle, and the experimental life. Princeton University Press, Princeton

Shelley PB (1826) Miscellaneous and posthumous poems of Percy Bysshe Shelley. W. Benbow, London

Sheppard E (2011) Geography, nature, and the question of development. Dialogues Hum Geogr 1(1):46–75

Singh J (1998) My trysts with the projects Bhakra and Beas. Uppal Publishing House, New Delhi

Tuan YF (1968) The hydrologic cycle and the wisdom of God: a theme in geoteleology. University of Toronto Press, Toronto

Vörösmarty CJ, Meybeck M, Fekete B, Sharma K, Green P, Syvitski JPM (2003) Anthropogenic sediment retention: major global impact from registered river impoundments. Global Planet Change 39(1–2):169–190

WCD (World Commission on Dams) (2000) Dams and development. A new framework for decision-making. Earthscan, London

Withers CWJ (1995) Geography, natural history and the eighteenth-century enlightenment: putting the world in place. Hist Workshop J 39(1):137–164

Chapter 3
Tibetan Water to Save China?

Chinese Discourse on the Great Western Route

Miriam Seeger

Abstract In recent Chinese political rhetoric concerning the promotion of the so-called Great Western Route of the South-North Water Diversion, participants in the discourse shape their arguments to meet both the demands of the internal debate within China as well as the government's agenda, to strengthen their position. This is achieved by selectively appropriating those elements of foreign as well as (reinterpreted) traditional Chinese knowledge about water diversion that fortify the respective positions in the internal debate. In their arguments, the discourse contributors stress not only steady economic development that relies strongly on sufficient water resources as prerequisite for social stability and national integrity, but they also connect economic power with the iconic and prestigious aspects of large-scale hydro-engineering projects as a means of strengthening China's powerful international position. By combining arguments found in foreign discourses and traditional Chinese models, the discourse participants aim to show that China can overcome her perceived weakness in contrast to Western countries and become a stronger and more 'modern' society. The domination and transformation of nature through large-scale water diversion is presented as the only means to save China.

Keywords Water diversion • China • Great Western Route • Political rhetoric • Mankind-nature relationship

China's water resources are unevenly distributed. Whereas North China experiences an increasing water shortage, resulting in environmental deterioration due to a combination of scarce resources and overexploitation, in South China large amounts of water either flow away unutilised or cause floods and disasters. This is often

M. Seeger, M.A. (✉)
Institute of Chinese Studies, Heidelberg University, Akademiestr. 4-8, 69117 Heidelberg, Germany
e-mail: miriam.seeger@zo.uni-heidelberg.de

M. Nüsser (ed.), *Large Dams in Asia*, Advances in Asian Human-Environmental Research, DOI 10.1007/978-94-007-2798-4_3,

© Springer Science+Business Media Dordrecht 2014

given as the reason why a South-North Water Diversion Scheme (*nanshui beidiao gongcheng*)[1] is necessary. Putting such a scheme into place would lead to a large-scale transformation of nature by mankind, although recent official rhetoric perpetuated by the Chinese government stresses a "harmonious coexistence of mankind with nature" (*ren yu ziran hexie xiangchu*)[2] (Zhongguo Gongchandang 2006).[3]

Thus, there seems to be a contradiction between the official rhetoric and large-scale hydro-engineering projects that aim at improving the natural world, thereby reflecting modernity's quest of transforming and dominating nature.

From a 'modernist' point of view, characterised by a technocratic mindset,[4] rationalisation of nature entails the assumption that – for example – river water is only a useful natural resource if controlled, evenly distributed, both spatially

---

[1]The idea of South-North Water Diversion is normally ascribed to Mao Zedong who already pointed to China's unevenly distributed water resources in 1952 at an inspection tour at the Yellow River: 'There is a lot of water in the South and little in the North, if only it were possible to borrow some [from the South]' (Wang 1957, p. 3; cf. Kao and Leung 1986, p. 301). The same argument is repeated as reason for accomplishing the official South-North Water Diversion Scheme (Qian and Zhang 2001, p. 264; see also RMRB 2003) as well as the additionally promoted Great Western Route (Deng et al. 1999, pp. 103ff; N. N. and Guo 1998a, p. 11).

[2]Official rhetoric of a 'harmonious coexistence of mankind with nature' is part of the larger strategy of building a 'socialist harmonious society' (*shehui zhuyi hexie shehui*) that has been officially approved in October 2006 at the Sixth Plenum of the 16th Chinese Communist Party Central Committee (henceforth CCP CC) with the long-term aim of securing China's sustainable development by closing the ever-increasing gap between the coastal region and the hinterland by means of raising living standard and production. The resolution of this plenum particularly stresses that a socialist harmonious society among many other elements should follow the 'overall requirement of a democratic rule of law, equity and justice, a honest and caring [society], [a society] full of vigour, a stable and orderly [society], [and a society in which] mankind and nature live in harmonious coexistence' (Zhongguo Gongchandang 2006). The term has already been mentioned in the resolution of the Fourth Plenum of the 16th CCP CC (Zhongguo Gongchandang 2004).

[3]All internet sources cited in this chapter are archived at DACHS – Digital Archive for Chinese Studies at the Institute for Chinese Studies, Heidelberg University; please visit the DACHS page at http://www.uni-heidelberg.de/fakultaeten/philosophie/zo/sinologie/digital_resources/dachs/citation_en.html.

[4]The term 'technocratic mindset' metaphorically draws on the assumption that the thinking, and hence acting, of a person is influenced by the 'modernist' agenda of rationality, pragmatism and a belief in progress and modernisation. The world, therefore, is not understood as an outcome of social processes, but rather as rational and technical processes that can be controlled. Large-scale technological solutions are favoured for solving social problems. Technocratic ideas were first popular in early twentieth century in the USA and are expressed, for example, in the American economist and sociologist Thorstein Veblen's book *Engineers and the Price System* (1921), in which he states that engineers, scientists and technicians are the best equipped to govern a country in the modern and industrialised age because of their rational, pragmatic and effective planning and thinking. Concerning the Chinese leadership since the 1980s, in comparison to their predecessors, many members of the Politburo of the third and fourth generation fit into the category of highly skilled technocrats with 'technical education, professional occupation and leadership position' (Li and White 1990, p. 22). Recent examples are Li Peng and Hu Jintao as hydraulic engineers, Jiang Zemin as electric engineer and Wen Jiabao as geologist.

and temporally, as well as extensively used; otherwise it is perceived as a waste or a threat.[5] Moreover, water diversion aims to provide a basis for even and stable economic development, and hence to further social stability and self-sufficiency. When examining large-scale hydro-engineering projects, one general feature of projects such as the Three Gorges Dam or the South-North Water Diversion Scheme is their function as icons of national pride. They are expected to highlight China's position of power in the international arena.

The South-North Water Diversion Scheme, developed since the 1950s,[6] has been announced by the State Council as late as 2002 as part of the larger strategy of Great Western Development (*Xibu da kaifa*).[7] It aims at improving China's water resources by providing water from the Yangzi River basin along three routes – the Eastern, Middle and Western Route – to the highly industrialised, arid Yellow River basin[8] (see Fig. 3.1). Water diversion projects include the construction of canals, tunnels and large dams.[9] Among other functions, such as generation of power and flood control, dams serve as a buffer, providing a steady flow of water into the receiving river basin.[10]

---

[5] Although flood myths like Great Yu who tamed the floods express an aspiration for an 'ordered world out of chaos' (Lewis 2006, p. 1) with a distinct division of rivers and the surrounding land and depict the untamed nature as a threat for human existence, in 'premodern' China mankind was still perceived as one part of nature. The assumption here is that with the appropriation of 'modernist' ideas the underlying thought patterns of mankind-nature relationship changed to perceive both as separate entities and nature as a mere provider of resources that should further human development.

[6] Institutions involved in research for the South-North Water Diversion Scheme in the 1950s had been the Ministry of Water Resources and the Yellow River Conservancy Commission.

[7] The Great Western Development Strategy is a policy that started in 2000. It was implemented in order to help the great Western parts of China to catch up with the industrialised coastal regions. The policy is largely based on infrastructural projects. For further information to this policy, see note 24 below.

[8] Three routes of the South-North Water Diversion Scheme have been officially approved. The Eastern Route along the old Beijing-Hangzhou Grand Canal and the Middle Route starting from the Danjiangkou Reservoir at the Han River leading northwards to Beijing are already under construction and are supposed to be completed in 2012 and 2014. The most technically challenging is the Western Route, which will tap the headwaters of the Yangzi River. For this route, feasibility studies are under way and completion is scheduled for 2050.

[9] Water diversion schemes cause similar social and environmental problems as large dams. Social problems include forced relocation of people, destruction of social structures and power struggles about access to water resources. Environmental problems include the transformation of a fluvial watercourse into a stagnant water body, hindrance or prevention of fish migration and large construction works in seismically active regions. For a much more detailed discussion of dams and diversion schemes in the Himalayas, see Pomeranz' *The Great Himalayan Watershed* (2009).

[10] During construction of the Middle Route, the Danjiangkou dam has been increased in order to enlarge the reservoir's volume. It might even be connected with the reservoir of the Three Gorges dam at the Yangzi River in order to increase water flow. If the Western Route is realised, a series of dams will be built on the tributaries of the Yangzi River. Each river will be dammed in order to create a reservoir from which the tunnels and canals will be fed. In the high and remote area of the Himalayas, this is a rather ambitious project.

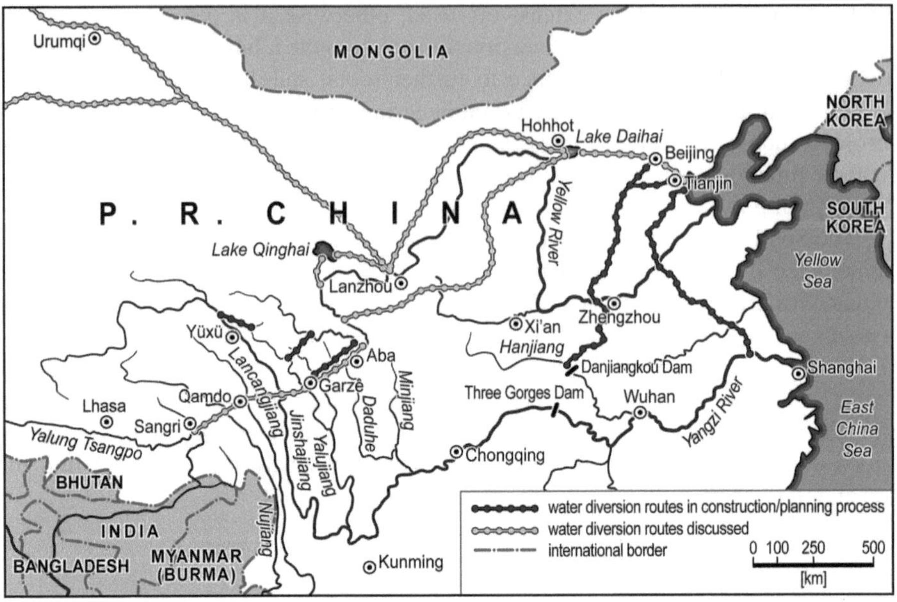

Fig. 3.1 Map of the different water diversion routes in China that are officially planned (Western Route) and currently under construction (Eastern and Middle Routes) as well as proposed by nongovernmental discourse participants (Great Western Route) (Seeger 2013, p. 182. Note: The courses of the official Eastern, Middle and Western Routes are drawn on the basis of Qian Zhengying and Zhang Guangdou's *Comprehensive Report on the Strategic Study on Sustainable Development of China's Water Resources* (2001, appendix, maps 14–16). The courses of the Great Western Route are drawn on the basis of rather sketchy maps in Li Ling's *Save China Through Water From Tibet* (2006, pp. 34, 71, 245). Since the Great Western Route is only proposed, this map depicts one possible course. The aim of this map is not to represent the different routes in detail, but to depict their different scales. I would like to thank the cartographer Nils Harm (Dipl.-Ing. (FH)) at the South Asia Institute's Department of Geography in Heidelberg for helping me draw up this map.)

Although the official water diversion project entails large-scale transformation of nature and has a social impact of great magnitude, it is legitimised by governmental institutions by their claim that it is an inevitable step towards securing China's sustainable development (Qian and Zhang 2001, p. 293). The project has been criticised by academics and activists of environmental and social organisations who question the benefit of such major projects in relation to their costs (see e.g. Fan 2007). On the other hand, it is also criticised as not being effective enough by a group of senior nongovernmental and younger academic discourse participants with a highly patriotic stance.

This latter position, which aims at even greater measures than those proposed by the government, is the topic of this chapter. It states that while the Eastern Route will divert expensive water of poor quality to Tianjin, the estimated amount of diverted water from the other two routes will not meet the demand of North China's large cities and the growing industry and agriculture in the next decade

(Guo 1998, pp. 4f; Shuili ziyuan diaopei 1999). The nongovernmental and academic key players predict that the Yangzi River will run short of water in 2020 (N. N. and Guo 1998a, p. 11; Zhongguo shehui 1998, p. 35); they therefore propose that China should capture not only the headwater of the Yangzi River (Jinshajiang, Yalongjiang and Daduhe) as the planned official Western Route will but also of the Tibetan rivers Yarlung Tsangpo (also known as Yalu Zangbu or Brahmaputra), Nujiang (or Salween) and Lancangjiang (or Mekong) (see Fig. 3.1). Instead of depending on water diverted from the Yangzi River, this so-called Great Western Route (*da xixian*)[11] would divert 200.6 billion cubic metres of water annually from the Tibetan Plateau – in their opinion a seemingly inexhaustible 'large natural water reservoir' (N. N. and Guo 1998a, p. 11) – to the Yellow River basin. It would bring the water to Beijing and Tianjin at the east coast as well as to the wide Northwestern desert areas along natural river beds and artificial canals, in order to transform the deserts into fertile agricultural lands.

The Great Western Route is praised by different discourse contributors not only as a panacea against North China's desertification problem but also as a strategy for saving China, strengthening her international position and thus improving her self-perception. These nongovernmental and academic discourse participants in favour of the so-called Great Western Route, some of whom will be introduced in more detail below, not only include in their arguments aspects of recent Chinese discursive formations such as an increasing regional development gap, social stability and self-sufficiency in terms of providing sustenance for the Chinese populace but also draw on traditional discursive formations such as effective river control as a sign

[11]There are different proposals concerning the starting point as well as the magnitude and construction process of the Great Western Route. Deng et al. (1999, ch. 9) specify some routes of importance: These include Guo Kai's proposal, which is entitled Shuo-Tian Canal (*Shuo-Tian yunhe*), because it connects the (not yet existent) Shuomatan Reservoir upstream of the Tsangpo's great bend with Tianjin at China's east coast (N. N. and Guo 1998a; Guo 1998). The three-step proposal by the Yellow River Conservancy Commission aims first to construct the so-called Small Western Route, thereby capturing the Yangzi River's headwaters; second, to extend the project to divert water from the Lancangjiang and Nujiang, and third, to dam the Yarlung Tsangpo's water (Deng et al. 1999, pp. 234ff). A further proposal is by the famous hydro-engineer Lin Yishan of the Yangzi River Conservancy Commission, who was deeply involved in the planning of the Gezhouba and the Three Gorges Dam at the Yangzi River. Lin's proposal does not include the Tsangpo (Deng et al. 1999, pp. 232ff), but in the future, when the demand for water is even greater and prices are cheaper, it could be extended to capture the Tsangpo as well (Cantian shuili ziyuan 2000). Deng et al. (1999) do not mention the proposal by Huang Wanli, the famous hydro-engineer of Qinghua University, who throughout his academic life opposed the damming of the main courses of large rivers, such as the Yellow and Yangzi rivers. His route would start at the Tsangpo's great bend (Li 2006, pp. 183ff; Zhao 2001, p. 344; Fan 2007, p. 71). A completely different Shuo-Tian Canal had been proposed by the authors of the *Collected Texts of the Shuo-Tian Canal* of the Geography Institute of the Chinese Academy of Sciences' State Planning Commission, starting in Shuo County in Shanxi province; it would divert the Yellow River's water eastwards to Tianjin (Wang et al. 1992, pp. 4f). This route allegedly refers to Guo Kai's first proposal of a Shuo-Tian Canal (Fan 2007, p. 71; Li 2006, p. 153).

for a prosperous government.[12] Moreover, they also take foreign water diversion projects – the Tennessee Valley Authority, the Rhine-Main-Danube Canal and the Siberian Rivers Reversal Project – as models for strengthening internal arguments of the political debate such as economic prosperity, strengthening of self-esteem as well as international power and national integrity.[13] This chapter uses an approach of analysing official discourse to reveal the process of how the proponents of a Great Western Route incorporate arguments of foreign as well as internal Chinese discourses into their own arguments. The analysis of these foreign water diversion projects suggests that the origin and the underlying original debates are of minor significance; rather, it is important to look at the particular function these models serve, what they represent and how they are integrated and domesticated into the internal discourse. By pointing to such foreign models, one could assume that the discourse contributors want to demonstrate how China learns from the 'West' and how modernisation can best be attained. This, however, would be a biased – or asymmetric – evaluation, that would view Western achievements and experiences as more valuable than those of Chinese origin. This chapter shows neither whether the discourse contributors have transferred the knowledge about these projects correctly nor whether their ideas have been influenced by these foreign ideas and led to the adoption of them. It rather aims at depicting how the appropriation of discourses proceeds. The various participants in the debate integrate those elements of the foreign discourses into their own arguments that strengthen their authoritative position within the internal debate, while excluding other elements in order to legitimise the authoritative position of the own arguments. I claim that the discourse participants who promote a Great Western Route shape their arguments to meet the government's agenda in order to strengthen their position in the debate. I will try to prove this assumption by showing that they draw strongly on topics that invoke the Chinese leadership's concern.

Further, they combine aspects of foreign as well as newly interpreted traditional Chinese models in order to show that China can overcome her perceived humiliation and weakness in contrast to Western societies with the goal of becoming a strong and 'modern' society. To reach this goal, the discourse contributors present the Great Western Route as the only means to save China.

By including materials from various popular and official sources, the position of proponents of the Great Western Route is represented by popular media and popular scientific sources, whereas official media and scientific sources represent the governmental position. To demonstrate the position of the advocators of a

---

[12]An increasing regional development gap, social stability and autarky are topics discussed by Deng et al. (1999, pp. 106ff), Lin Yishan (Cantian shuili ziyuan 2000) and Yuan Jiazu (1998; Cantian shuili ziyuan 1999). The traditional argument of effective river control as sign for a prosperous government is stressed by Guo Kai (N. N. and Guo 1998b). These topics are discussed in this chapter further below.

[13]Usage of foreign water diversion projects as models in order to strengthen internal discourse arguments can be found in an interview with Guo Kai (N. N. and Guo 1998b).

Great Western Route, I draw on articles by, and interviews with, Guo Kai,[14] a self-fashioning senior as well as a nongovernmental and nonscientific voice in the discourse, who reportedly was the first to come up with the idea of diverting water from the Yarlung Tsangpo to North China.[15] Except for one article published in the journal for high-ranking cadres, *Internal Consultative Readings*[16] (cf. Guo 1998), this text corpus was published in the academic journal *Trends in Contemporary Thought*[17] (cf. Guo 2001; N. N. and Guo 1998a, b, c) that is associated with

---

[14]The story of how Guo Kai came upon the idea of the Great Western Route is very interesting. During the so-called Cultural Revolution in the 1960s and 1970s, Guo, reportedly 'groundless accused', was held captive alone in a basement room for 22 months, where he by chance gained possession of an English edition of a book by the early twentieth century geologist Weng Wenhao about Chinese geography, which stimulated him to think about diverting Tibetan water to North China (Li 2006, p. 119). While reading Weng's book, he was able to draw on two earlier life experiences: During underground political work in the late 1940s, Guo attended a course of lectures about water conservancy at Qinghua University, as well as in the early years of the People's Republic, while attending a People's Liberation Army armed working unit for 'liberating' the far Southern and Western regions, Guo gained an impression of the climatic and environmental conditions in these areas. After the Cultural Revolution (1966–1976), inspired by Weng's book, he began consulting experts of the Ministry of Water Resources and the Chinese Academy of Sciences and made extensive field surveys along the route he proposed. Guo tried hard to arouse interest for his idea and was particularly successful among – still influential – senior cadres and military generals. Not having any official background, he is characterised by the Chinese media as a 'popular water resources expert' or a 'person at the margin of water conservancy' (N. N. and Guo 2006). His authority in terms of water resources management is based on his family history. Guo reportedly is a descendant of the Yuan Dynasty (1271–1368) water resources scholar Guo Shoujing, which roused his interest in rivers and water from his early youth (Li 2006, pp. 116ff). Since Li Ling's book *Save China Through Water From Tibet* (2006), the source for Guo Kai's biographic information, is mainly based on interviews, it seems likely that this claim is part of Guo's self-promoting strategy of strengthening his authority in the discourse.

[15]A translator's note to Fan Xiao's article in *The China Environment Yearbook* contends that the idea and study concerning the Great Western Route dates back to the 1950s (Fan 2008, p. 191, note 7).

[16]*Internal Consultative Readings*, a Chinese language journal for high-ranking cadres in government, party and military, is published since 1990 by the People's Daily editorial. Between 1990 and 1991, it was published irregularly; since 1992 it is published weekly. The journal deals with important Chinese and foreign political, economic, cultural, diplomatic, military and societal issues that are judged as unsuitable for appearance in public newspapers. Since *Internal Consultative Readings* is readable in public libraries such as the Shanghai Library (issues from 1995 until the most recent issues), it seems it is not longer classified.

[17]*Trends in Contemporary Thought*, a Chinese language journal published from 1990 to 2004, is organised by the Association of National History (*Guoshi xuehui*), a direct subordinate of the Chinese Academy of Social Sciences' (CASS) Institute of Contemporary China Studies. Chief editor of the journal is the CASS member and well-schooled philosopher Duan Ruofei, who formerly worked at the Secretariat of the CCP CC (Eckholm 1999; Feng 1999). *Trends in Contemporary Thought* has allegedly functioned as a mouthpiece for conservative leftist ideas with strong criticism of the government's reforms and concerns that China is turning into a capitalist society (Feng 1999). Since 2005, the journal *Trends in Contemporary Thought* is succeeded by *China-Today Forum*, published by the China Association of Policy Science (*Zhongguo zhengce kexue yanjiuhui*).

the Chinese Academy of Social Sciences (henceforth CASS). Further sources include scientific and popular scientific texts associated with the Beijing-based CASS's Economy and Culture Research Center, represented by younger academic contributors such as its director, professor of economy Deng Yingtao, and the economist Wang Xiaoqiang.[18] The Economy and Culture Research Center, by drawing on the official Great Western Development Strategy in connection with water resources management, has a special interest in 'rebuilding China', as the title of a book publication by Deng et al. (1999) illustrates. The text corpus also includes articles published in the Economy and Culture Research Center's newsletters (cf. Zhongguo shehui kexueyuan n.d.). Material for analysing the government's response to the discourse and critique of the Great Western Route project consists of governmental scientific and media sources. The material includes a publication by the Chinese Academy of Engineers, the *Comprehensive report on the strategic study on sustainable development of China's water resources and different position papers* (Qian and Zhang 2001) and articles in academic journals such as the Ministry of Water Resources' publication *China Water Resources* (cf. Wang 2004).

In Western academic circles, contrary to the project of the Three Gorges Dam, the topic of the Great Western Route or water diversion from South to North in general has not been discussed intensely. Most of the rather few studies about water diversion from Himalayan rivers and large dams in the Himalayas look from either a regional or a trans-boundary river perspective. Consequently they discuss not only China's but also India's and South Asia's interests and concerns,[19] represent the perspective of Tibetan people[20] and analyse the environmental impacts of large-

[18]The interest of these younger urban-educated academics in the topic of rural and regional economic development was aroused during the Cultural Revolution, when they were sent to the countryside (Fewsmith 1994, p. 34; Liu 2009; Wang 1999). During the 1980s, Wang Xiaoqiang and Deng Yingtao were core members of the think tank China Rural Development Research Group (*Zhongguo nongcun fazhan wenti yanjiuzu*), subordinate of the CASS's Agriculture Institute, which aimed for fundamental rural reform (Fewsmith 1994, pp. 34ff; Liu 2009).

[19]Studies with a regional approach include *The great Himalayan watershed: agrarian crisis, mega-dams and the environment* by Kenneth Pomeranz (2009), whose research focus is on Chinese modern social and economic history. It illustrates that China is by no means the only, but the most successful country depending on Himalayan water that solves her water problems at the expense of other countries. Vaclav Smil (2009), whose studies focus on China's environment, published *Finding mutual interests in nature* in the popular journal *Far Eastern Economic Review*, in which he draws on historical models for China's increasing pace and scope of constructing mega dams and water diversion projects by taking into account the concerns of South Asia and India. Claude Arpi (2008), a journalist and observer of the relationship between India, China and Tibet, stresses the importance of river water treaties in the article *Himalayan rivers: geopolitics and strategic perspectives*, published in the *Indian Defence Review*.

[20]The perspective of the Tibetan people is represented by two reports with an academic approach, but with a stronger intention of political influence. These are *HydroLogic: water for human development* by Tashi Tsering (2002), an environmental researcher, whose focus is on the Western Himalaya, published in association with the Tibet Justice Center, which advocates human rights and self-determination for the Tibetan people, and one chapter of *Tibet: a human development*

scale water diversion projects in general.[21] Since the formation of internal debates in China has not been reviewed, this chapter aims at analysing the formation of such discourses, focusing in particular on the appropriation of foreign as well as internal Chinese debates in order to reveal the discourse contributors' strategies of strengthening their own position.

First, I will show how the participants in the discourse shape their arguments to meet the demands of internal debate in China with the aim of strengthening their own project. This will be exemplified on the basis of the Tennessee Valley Authority and the Rhine-Main-Danube Canal. In the analysed discourse, the former is presented as a model for boosting economic power, whereas the latter is interpreted as a model for raising national self-esteem. In addition to citing these foreign models, the proponents of the Great Western Route also draw on the Chinese traditional knowledge of the Grand Canal, which was built in imperial times for the transportation of tribute grain from the Southern economic centre to the capital in the North.[22]

The second section of this chapter will analyse issues of social stability on the basis of the Siberian Rivers Reversal Project. The failure of this project is interpreted to be a direct cause for the break-up of the Soviet Union, thus serving as a warning of the consequences of neglecting proper water diversion for large, multiethnic countries like China. This topic also encompasses the traditional Chinese concept of taking effective river control as a sign of a benevolent ruler.

3.1 Models for Economic Prosperity and International Power

The 'TVA idea', as depicted by one of the Tennessee Valley Authority's early directors, David Lilienthal, in his book *TVA: Democracy on the March* (1944), is often presented as a multipurpose approach of regarding the whole river basin as one unit. It not only aims to generate electricity, improve flood control and

---

and environment report (Environment and Development Desk 2007), compiled by the Tibetan Government in Exile.

[21]In an article published in the *Environment Yearbook* edited by the environmental nongovernmental organisation Friends of Nature (*Ziran zhi you*) that is published in an English translation by Yang Dongping as well, the geologist Fan Xiao highlights the environmental impacts of large-scale construction projects, among these are the proposals for the Great Western Route and the official Western Route of the South-North Water Diversion Scheme. He analyses impacts of such construction works on the natural environment due to geological conditions, but also impacts on the societal environment and international relations (Fan 2007, 2008).

[22]The first sections of the Grand Canal, which was closely connected to the hydraulic system of the Yellow River, were built in the fifth century BCE. Yet the different sections were connected for the first time in the Sui Dynasty (581–618) and extended as well as reconstructed in the Ming Dynasty (1368–1644). For an introduction to the Grand Canal, see e.g. Leonard's *Controlling From Afar* (1996, ch. 1), Elvin's *The Retreat of the Elephants* (2004, ch. 6) and Chi Ch'ao-ting's *Key Economic Areas in Chinese History* (1963).

navigation, provide water for irrigation and stimulate recreation but also aims to increase economic development and employment. The TVA furthermore is based on sociopolitical elements that include the propagation of a democratic ethos and bureaucratic decentralisation and encouraging decolonisation by transmitting the idea of development and modernisation (Lilienthal 1944; see also Ekbladh 2001; Russel 1949, pp. 43–51).

In Guo Kai's argument for the Great Western Route, he stresses the TVA as a model for how the USA became the world's biggest economic power. He states that the USA could never have been able to gain the leading position in the world based solely on their success in the Second World War. Guo stresses the importance of the TVA in developing the Tennessee Valley in the 1930s and the subsequent impulse for the development and modernisation of the USA (N. N. and Guo 1998b, p. 3). Zhang Zhengbin, a research fellow at the Center for Agricultural Resources of the Chinese Academy of Sciences (henceforth CAS), sees the successful transformation of the USA's arid Southwest and the development of metropolises in desert areas due to long-distance water diversion as an evidence for the Great Western Route's feasibility, and significance for China's development. Parallel to the development in the USA, he argues, large-scale water diversion in China's northwest will strongly improve the living standard as well as state power (Zhang et al. 2008).

Interestingly, the presentation of the TVA and US achievements in modernisation due to water diversion is only valued on the basis of economic progress. Instead of including sociopolitical elements like furthering democracy and decentralisation when translating the TVA project into the Chinese context, it is narrowed down to the aspect of gaining the status of the world's biggest economic power and its value for modernisation. However, the TVA is a well-known project in Chinese academic circles and has been discussed since the republican era, among other things as way of establishing a Yangzi Valley Authority and its value for implementing democracy in China (cf. Mao 1947). It is therefore not unimaginable that the TVA could function as a surrogate discourse to indirectly promote democracy. Nevertheless, in the context of the Great Western Route, this seems implausible, because aspects like democracy do not meet the discourse contributors' interests. From the analysed sources, it is impossible to discern whether the potentially associated sociopolitical elements prevail as the TVA's dominant attributes in the minds of the discourse's recipients. However, the TVA model meets the government's aim of modernisation and economic development, as well as the subsequent gain of power in the international arena very well.

In addition to highlighting the USA's gain in power through large-scale water diversion, the discourse participants in favour of a Great Western Route point to the Rhine-Main-Danube Canal that serves as an even stronger model for strengthening a country's self-perception and international power position. The Rhine-Main-Danube Canal in Germany, providing a navigable route between the Rhine and Danube basins, was completed in 1992. Guo Kai's depiction of this canal is based on the assumption not only that it provides economic benefits for agriculture, industry and tourism but also that the project was accomplished with some idealistic goals in mind: to realise the eighth century German Empire's dream, represented by the unfinished first attempt to connect the Rhine and Danube basins by emperor Charlemagne (747/748–814), the so-called *Fossa Carolina* or *Karlsgraben*, and, more recently, to strengthen the Germans' spirit after the defeat in the Second World War. He states that in politics and ideology, the canal produces unimaginable 'social effects' and causes unimaginable 'international repercussions' (N. N. and Guo 1998b, p. 3). According to this argumentation, the canal functions as a means of restrengthening the Germans' national identity.

Determining the importance of patriotic or nationalistic feelings in the German debate on the Rhine-Main-Danube Canal is beyond the scope of this chapter.[23] The translation of the Rhine-Main-Danube Canal into the Chinese context as a model for strengthening self-perception and national pride, however, shows a clear analogy to the internal Chinese discourse: China's aim of regaining a powerful position as a respected nation has been a fundamental issue of every Chinese government. Due to the contact with Western modern technology and military power in the late imperial period, the Chinese perceived themselves as a weak and stagnant nation. In resemblance to the traditional concept of the Chinese emperor as the spiritual centre of the (Chinese) world, whose power emanates from the imperial centre to the peripheral areas due to his benevolence, a strong motive of every Chinese government is to re-establish China as the world's leading power (Pietz and Giordano 2009, p. 118). The recent strategy of Great Western Development is such a strategy that by solving internal problems that have arisen due to economic reform,[24]

[23]For an analysis of a German newspaper discourse on the Rhine-Main-Danube Canal in the 1980s, see Wirth's article on the sense and nonsense of that navigation route (1983). Wirth shows that in the German discourse, the arguments in favour of the Rhine-Main-Danube Canal show five kinds of arguments: to create a continuous waterway from the North Sea to the Black Sea; the stop of the canal's construction beyond Nuremberg would be a waste of investment; the canal functions as a lifeline for Austria's economy; the canal as indispensable for Franconia's water and energy supply; and the canal as a stimulus for recreation and tourism. As visible from Wirth's analysis, there is no (nationalistic or patriotic) discourse of strengthening a German identity. However, the discourse includes elements of strengthening the region's – i.e. Bavaria's – identity as Germany's number one industrial and business location and as centre of the European market, especially after the Soviet Union's disintegration during the late 1980s and early 1990s (Wirth 1983, pp. 45ff). In this sense, it is claimed that the canal should have a 'unifying force for different nations' (völkerverbindende Kraft) (1983, p. 49) and that Nuremberg 'one day will be *the centre of European inland navigation transport*' (1983, pp. 53f, originally quoted in *Nürnberger Nachrichten*, 16–17 September 1972).

[24]The Great Western Development Strategy is an amendment to the policy of economic reform, implemented in the late 1970s, which promotes that some regions should become rich first. Only a decade later, Deng Xiaoping saw the need to justify his policy by formulating a two-stage regional development theory: first, the coastal regions should attain central support; once they achieved a sufficient level of development, the focus should be shifted to the interior regions (Tian 2001, pp. 78ff). The main academic discourse during the 1980s assumed that the development disparity between the coastal region and the hinterland would be reduced by a gradual diffuse of the coastal region's technologies, economic progress and wealth into the hinterland. The growing gap between these two regions, however, did not decrease, but in fact increased even more, as opponents of this 'trickle down' theory showed. In the course of discussion of regional development disparities during the 1990s, the intellectual justification for the Great Western Development Strategy had been set (Holbig 2004, pp. 336ff).

especially an increasing development gap between coastal areas and the hinterland, aims at improving China's international position. It not only is a nation-building strategy that encompasses the aim of social homogenisation as well as social and political stability in non-Han areas,[25] which is officially referred to as 'national' or 'ethnic unity' (*minzu tuanjie*) but also is an effort in state-building with the aim of strengthening China's governmental, infrastructural and economic development on the whole (Goodman 2004, pp. 324ff).

By interpreting the Rhine-Main-Danube Canal in the context of regaining Germany's national identity and power after defeat in war and emphasising the USA's achievements in transforming deserts and supplying water to metropolises, these foreign examples become strong and obvious models for China. During the translation process of both foreign models, transformation of nature emerges as an instrument to gain sociopolitical and economic achievements. Nature is valued according to output figures, and thereby the existence of different landscape characteristics is opposed. This is expressed by Guo Kai when highlighting the transformation of North China's deserts into 'oases'[26] (2001, p. 54) or the construction of a second 'Jiangnan' (2001, pp. 55f), the area south of the Yangzi river delta, with 'ever green hills and ever flowing blue rivers [lit. green rivers, because in Chinese green is the ideal colour of water in the natural context]' (2001, p. 56). He points to the Great Western Development Strategy's advantage, that through the improved environment, China can make a 'great economic leap' to 'exceed Japan and catch up with the USA' and, consequently, become the 'world's first economic power' (2001, p. 56). Thereby, Guo recalls ideas from the Great Leap Forward's rhetoric of the late 1950s to outpace the 'West' in one great effort (cf. Schoenhals 1987, pp. 1ff; MacFarquhar 1983, pp. 15ff). The argument of improving the Northwestern environment obviously goes along with the official strategy of opening-up the Western parts by highlighting the importance of the hinterland's development in furthering all of China's development and hence improving China's international position.

Taking the government's rhetoric into account, in terms of terminology, we can see a disparity between the positions of official representatives and especially the senior advocators of a Great Western Route. Guo Kai's adherence to struggling against nature (2001, p. 52) and opposing the existence of different landscapes (2001, p. 55) is criticised by governmental representatives of the Ministry of Water Resources.[27] If we look beyond the terminology at the underlying perceptions of

---

[25]The Han population of China is the largest ethnic group. In addition to the Han majority, China officially recognises 55 ethnic groups, the so-called ethnic or national minorities.

[26]The reference to creating 'oases' seems to be a misleading metaphor. It evokes an image of small, green, moist and fertile islands in the midst of the scarce and hostile desert, which like a 'fata morgana' may only be an illusion. At least an oasis cannot encompass the whole desert. The proposals, however, depict that the Northwestern parts at large shall be transformed into fertile agricultural fields.

[27]Qian Zhengying, the former Minister of Water Resources, respectively, Water Resources and Electric Power (in office 1974–1988), refutes the metaphor of constructing a second 'Jiangnan'

nature, it becomes clear that both positions express a technocratic mindset and are deeply influenced by modernity's distinction of mankind and nature. Both harbour the notion that nature has to serve society as a provider of resources for satisfying humanity's needs. The government's concept of 'harmonious coexistence of mankind with nature', literally understood, reflects the existence of two systems – humanity and the physical world around it – that are characterised by mutual noninterference.[28] However, it is based on the term 'sustainable development' that is defined in *China's sustainable development agenda in the early 21st century* as treating mankind as the basis and economic development as the core (Renmin ribao she 2003), which means that this concept is anthropocentrically motivated. In China, similar to the World Commission on Environment and Development's definition (WCED 1987), the term 'sustainable development' incorporates the attitude that the exploitation of nature in the course of economic development should not be exaggerated, so that it does not jeopardise the existence of future (Chinese) generations.[29] It is a means of catching up with the industrialised countries, by pointing to China's achievements in fast but sustainable development. Although 'harmonious coexistence of mankind with nature' and ideas of transforming nature seem to be mutually exclusive, the official rhetoric merges them by postulating that nature has to be profitable and beneficial for humanity. This mindset is the intellectual foundation for attempts to transform nature such as even allocation of water resources through construction of the South-North Water Diversion Scheme (RMRB 2003). Instead of recognising the redirection of water and the transformation of Northern landscapes as a human interference into ecosystems, it is understood as an act of improving nature that is assumed to be imperfect. The official perception of the mankind-nature relationship, however, documents an attitude that resembles those expressed by the proponents of the Great Western Route.

In conclusion, the proponents of a Great Western Route appropriate topics of the official discourse into their own arguments. A recurrent theme is the alleged gain in national strength through development of the Western areas. In this context, the model of US water diversion and the TVA as well as the Rhine-Main-Danube Canal

---

in North China when calling it 'violating the laws of nature' (2001, p. 7). Another former Minister of Water Resources, Wang Shucheng (in office 1998–2007), distances himself from the idea of struggling against nature when saying that due to the historical lessons drawn from the excessive exploitation and destruction of nature in the Mao era, mankind today lives in 'harmonious coexistence with nature' (2004, p. 7).

[28]For an introduction to the slogan of a 'harmonious coexistence of mankind with nature', see also another paper of mine. There I discuss water diversion in China as seen between official rhetoric and the drive to dominate nature (Seeger 2013).

[29]The definition of 'sustainability', of course, is very diverse. This paper, however, focuses on the difference between sustainable development defined anthropocentrically such as by the World Commission on Environment and Development of 1987: 'development that meets the needs of the present without compromising the ability of future generations to meet their own needs' (WCED 1987), and defined ecocentrically: nature and environment are taken as the core point and mankind as only one inhabitant among many on earth; thus mankind has the responsibility to conserve the treasures of the earth (e.g. Shiva 2006).

provide strong arguments for the internal debate in China – at least when we look beyond the rhetoric and take the underlying assumptions into account, considering the way the models are translated into the Chinese context. Economic development and the international status associated with it are fundamental aims not only of the supporters of the Great Western Route but also of the Chinese government.

In his argumentation, Guo Kai not only points to the USA's international influence and its assumed connection to the TVA and the Rhine-Main-Danube Canal as a way of supposedly raising a country's national sense of identity but brings to further attention the traditional model of the Grand Canal as a type of knowledge exemplary of China's worldwide influence. He claims that the Beijing-Hangzhou Grand Canal functioned as a successful model of water conservancy and promoted social development that spread throughout the world. Inspired by the Chinese, various countries allegedly built canals (N. N. and Guo 1998b, pp. 2f); however, he does not support his claim with examples.[30] By not merely relying on foreign models, but on China's ancient achievements as well, Guo Kai strengthens the domestic patriotism and self-esteem in order to re-establish China's leading position as a benevolent and harmonious centre of the world. His argument is quite different from that of the modernisation discourse in the first half of the twentieth century, such as expressed by Sun Yatsen, who aimed at catching up with Western modernisation and clearly valued Western knowledge of water conservancy as advanced and superior to local Chinese river control knowledge (Sun 1929). Guo's argument, however, seems akin to the late imperial theory known as 'Chinese origin of Western science' (*Xi xue Zhong yuan*), which claims that diverse kinds of Western knowledge originated in China. It was a strategy to cope with the intensifying contact with Western knowledge and clearly aimed at strengthening the Chinese position of power that was perceived to be being eroded.[31] Guo Kai aims to re-establish China's assumed 'natural' position as centre of the world by pointing to the model of the TVA – or river control in general – for gaining an internationally powerful position and referring to China as preceding the 'West' in coming up with progressive ideas that spread throughout the world. In this fashion, Guo complements the patriotic model of China's long history, the Grand Canal, with the TVA as a model for catching up with the world's leading economic and political powers. The apparent asymmetry in knowledge is eased by pointing to the alleged 'Chinese origin of Western knowledge', thus strengthening China's national identity. Furthermore, successful river control is attributed with having a direct effect on a nation's self-

[30]After stating that the Chinese inspired many countries to build canals, Guo Kai points to the Soviet Union and the USA's achievements of cutting canals in their interior regions in the 1930s (N. N. and Guo 1998b, p. 3). If he values both countries' achievements as influenced by the Grand Canal of China, which dates back to late sixth and early seventh century, he somewhat broadly interprets the influence of a historical achievement.

[31]The theory of 'Chinese origin of Western science' is discussed, for example, by George Wong (1963) with the focus on its 'psychological effectiveness' in late Ming and early Qing (1644–1911) China. It was used by traditionalists 'to de-popularise and to minimise' the influence of Western scientific ideas and techniques that were introduced by the Jesuit missionaries (Wong 1963, p. 30).

esteem, substantiated by the model of the Rhine-Main-Danube Canal. Thus, China can become a strong and modern society by looking at successful foreign examples as well as by learning from her own historical knowledge.

The proponents of the Great Western Route shape their arguments to meet the government's concerns by integrating the TVA and the Rhine-Main-Danube Canal into the Great Western Route discourse as models for gaining regional economic prosperity and national strength and connecting these with the traditional model of the Grand Canal, depicted as an element of Chinese knowledge that precedes the Western achievements in water resources development, thus evoking a sense of patriotism. However, not only successful foreign examples serve as models in the debate on the Great Western Route. The abandonment of the Siberian Rivers Reversal Project, translated into the Chinese context, serves as a warning by evoking the fear of the nation's disintegration. On the basis of this foreign model and (reinterpreted) traditional ideas of effective river control as a sign of benevolent rule, the next part will discuss how proponents of the Great Western Route depict river control as a means of social stability and national unity in contemporary China.

3.2 The Threat of National Disintegration: A Warning

In contradiction to the two foreign examples above, the third example constitutes a warning rather than a model for China. Davydov's 1949 plan for a Siberian Rivers Reversal Project,[32] part of Stalin's 'Great Plan for the Transformation of Nature', would have reversed the flow of the rivers Yenisei, Ob and Irtysh. Instead of flowing into the Arctic Ocean 'unutilised', they would have been diverted to Central Asia for irrigation and supplementing the water volume of the Aral Sea. Davydov's argumentation bears a striking resemblance to the argument in favour of the Great Western Route, which highlights that only by supplying water, deserts will transform into oases: 'Give this [desert and semi-desert] land water and the deserts will come to life and new gardens will bloom in Central Asia and Kazakhstan' (1972, p. 606).

[32]Davydov introduced the project in 1949 by saying: 'We are building communism, we are transforming life on earth. [...] the Ob and Yenisey will empty into the Caspian Sea. [...] This is neither a joke nor a useless fancy. It is a problem of national importance with enormous significance for the development of the entire economy of our homeland' (Davydov 1972, pp. 603f). For more information about the Siberian Rivers Reversal Project, see for example, Duke's article *Seizing favours from nature: the rise and fall of Siberian river diversion* (2006). Therein he argues that the reason the Siberian Rivers Rerouting Project did not receive the central government's support can mainly be attributed to competing economic interests (Duke 2006, p. 5). Giese (1998) shows that early criticism on Davydov's gigantic plan was evoked by a concern for catastrophic influences on the water balance of the Western Siberian Plain. Later plans by Gidroproyekt in the 1960s were abandoned due to economic reasons: the discovery of oil and gas in the area that would have been inundated (Giese 1998, p. 94).

According to Guo Kai, Stalin's plan of a Siberia to Central Asia North-South Water Diversion Project – as he calls it parallel to the Chinese terminology – was abandoned when Khrushchev came to power in 1953 after Stalin's death. Guo blames the increasing desertification of the Central Asian region on Khrushchev's policy that forced land reclamation and cultivation of corn under the condition of large-scale water scarcity.[33] This policy caused a widening gap in the societal development between the Eastern and Western parts of the Soviet Union, furthered economic depletion in the Southern states of the Soviet Union with ethnic minority populations, led to large-scale population movements from Central Asian to Soviet European regions and finally led to increased social and ethnic conflicts (N. N. and Guo 1998b, p. 4; Yuan 1998, p. 9). Guo points to that which Yuan Jiazu, professor of forestry and a supporter of the Great Western Route, explicitly states: Khrushchev's abandonment of the Siberian Rivers Reversal Project 'certainly had an influence on the Soviet Union's disintegration' (Yuan 1998, p. 9).

Whether Khrushchev's change in water diversion politics had influenced the Soviet Union's disintegration or not, it is important to note that for China, as a territorially large country with various local cultural formations, a possible disintegration of the state is one of the main concerns. Social equality and even economic development as well as social stability are fundamental justifications for water diversion in China. South-North Water Diversion and the government's policy of Great Western Development aim exactly at solving the problem of increasing development disparities between coastal areas and the hinterland. By supporting the official programme for developing the Western regions, the Great Western Route advocators Deng Yingtao and Wang Xiaoqiang with their concept of 'rebuilding China' (Deng et al. 1999, pp. 4f) stress the government's duty of providing social equality through evenly distributed natural resources and economic development. They support the – in comparison to the last 20 years of reform (Goodman 2004, pp. 318f) – higher degree of state intervention in the economic development of the interior regions. It seems that the supporters of the 'rebuild China' concept depict their strategy as the 'real' Great Western Development Program, because only through the Great Western Route, they argue, can a fundamental and sustainable development of the Northern region and China as a whole be achieved (Shui ziyuan diaopei 2000a, b; Wang 2009).

The widening developmental gap between coastal areas and the hinterland challenges the state's role as guarantor for even development and adds urgency to the issue of social stability. The famous hydro-engineer Lin Yishan of the Yangzi River Conservancy Commission, who also made recommendations on water diversion from the Himalayan Plateau,[34] states in an interview with Deng Yingtao

[33]Here Guo refers to the Virgin Land Program that started in 1954 under Khrushchev. The policy was intended to increase the Soviet Union's wheat production through reclamation of uncultivated lands in Central Asia and Kazakhstan. For more information on this policy, see McCauley's *Khrushchev and the Development of Soviet Agriculture* (1976).

[34]For Lin Yishan's proposal of a Great Western Route, see note 11 above.

that the increasing social inequality between the less developed peripheral areas with large non-Han population and the developed coastal regions with mainly Han-Chinese population might lead to ethnic conflicts that should be alleviated by a centrally organised 'mixed residence of [different] ethnic groups' (Cantian shuili ziyuan 2000), which means migration of Han-Chinese into the peripheral areas. Social instability can also be caused by China's growing population. Deng et al. argue in *Rebuild China* (1999) that the construction of a new economic centre in the Northwest through water diversion is the most effective method of solving the population issue in the long run. Other approaches, such as the relocation of people to the industrialised but already overpopulated Southeast and reallocation of food to Northwest China by risking an import dependency, they deem inferior (1999, pp. 106ff, 117f). The population issue coupled with China's desire for self-sufficiency in food production is an important topic for the Chinese government, which is stressed by Yuan Jiazu (1998, p. 6) when pointing to an October 1996 government's White Paper entitled *The grain issue in China*. This White Paper contends China's ability to feed her people self-sufficiently without having to rely on foreign imports (Zhonghua renmin gongheguo n.d.). Yuan's argument, however, is that China is bound to improve the Northern agricultural areas by means of water diversion, based on the Great Western Route, if the government wants to fulfil its desire for autarky (1998, p. 6).

Contrary to the 1950s, when a growing population was promoted to strengthen the labour force and, consequently, speed up the socialist construction (Shapiro 2002, ch. 1), the White Paper on the grain issue today depicts the government's strong concern about social instability caused by uncontrolled population growth as well as unrestricted internal migration. Governmental actions concerning ethnic minority areas such as development of Tibet or Southwest China are characterised by large-scale engineering projects. The Qinghai-Tibet Railway, for example, brings 'modernity' to Tibet, but also boosts migration of Han-Chinese to Tibet for economic reasons or because they are being sent by the government as officials and technical experts. Large dams in Yunnan province – and recently in Tibet as well – bring technology and industry; high qualified jobs, however, will not be assigned to domestic people, but to immigrants. In the same fashion as proposed by Lin Yishan, the government assists migration of Han-Chinese to the border areas as a means of creating a Chinese identity, which ought to encompass all 'Chinese people' (*Zhonghua minzu*) regardless of their ethnicity, through standardisation and homogenisation by sharing language and social customs as part of the nation-building strategy of Great Western Development (Goodman 2004, pp. 324ff).

In conclusion, the advocators of a Great Western Route appropriate topics of the normative discourse into their own arguments in favour of the Great Western Route, in order to strengthen their position of authority. Such topics encompass issues the government is concerned about, such as strengthening social unity and stability through development of the hinterland. By integrating the warning of the alleged consequences of the Siberian Rivers Reversal Project's abandonment, the advocators of a Great Western Route caution China about the risk of uncontrolled migration, ethnic conflicts and, consequently, the country's disintegration by failure to accomplish their proposed giant scheme. Whatever the reasons for the Soviet Union's disintegration were, the idea of depicting a large-scale water diversion project as a unifying factor for territorially large states is a very strong argument in the Chinese context.

The relationship between a stable government and control of rivers, such as demonstrated with the Soviet experiences, is a fundamental theme in the Chinese cultural consciousness, expressed in the traditional idea of the 'mandate of heaven', which means a celestially sanctioned rule as sign of legitimate government.[35] Guo Kai demonstrates that the state of a country's waterways is a symbol of its prosperity or decay with an old precept[36]: '[If one wants] to govern a country, [one] first has to govern the rivers; governing the rivers, then one can govern a country; that is why Gun failed and Yu succeeded'[37] (N. N. and Guo 1998b, p. 2). The importance of controlling the rivers for governing a state is already indicated in the Chinese classical text *Guanzi*[38] in the chapter *On appraising the terrain* (Guanzi, ch. 57, pp. 202a–205b). In this passage, Guan Zhong says to Duke Huan of Qi: 'Therefore, those who [desire to be] good at ruling their country first have to eliminate the five evils' (Guanzi, ch. 57.1, p. 202b; cf. Guan and Rickett 1985, p. 244). Water he specifies as the greatest of these evils; if the water is not well controlled, the society will come into disorder (Guanzi, ch. 18.2, p. 203a). The *Guanzi* does not refer to Great Yu in this context. Nevertheless the historical and mythological records hold many references to Yu as the great engineer ruler, who at the time of the great deluge had confined the wild-flowing rivers to their courses and founded the first historically documented dynasty.[39] Occurrences of natural disasters caused by

[35]The concept of the 'mandate of heaven' is an ancient Chinese philosophical idea that is based on the justly, virtuous and benevolent rule of the king. If the ruler becomes despotic, he loses his mandate. Natural calamities such as floods, droughts, plagues and earthquakes are signs of the mandate's loss, just as rebellions are legitimate means to establish a new rule.

[36]The research is based on the relevant Chinese ancient text databases with a combined keyword search. The result is that the sentence is not a quotation from an ancient Chinese text. The earliest source of this quotation I could locate is in the *Shenbao* of 10 April 1938, the first Chinese daily newspaper published in Shanghai from 1872 to 1949, in an article in honour of Li Yizhi, the so-called father of hydrology in China. This source, however, does not make a connection to Yu and Gun (Shenbao 1938).

[37]The legend of Gun and Yu is as follows: At a time when China was devastated by floods, Gun was ordered by the legendary king Yao to control the floods but failed in his task. Gun's method of controlling the rivers was to build dams and dykes. After Gun's execution, his son Yu, who was ordered by Yao's successor Shun, completed his father's task. Instead of damming the rivers, however, he dredged the river beds and cut new channels so that the water could drain off to the sea. For an introduction to the legends of Gun and Yu, see Anne Birrell's book on Chinese mythology (1993, pp. 146–159) or Mark E. Lewis book on flood myths in early China (2006).

[38]The *Guanzi* is written by Guan Zhong, a scholar and politician of the so-called Spring and Autumn period (770–476 BCE) in the seventh century BCE of the state of Qi. The *Guanzi* was translated and commentated by Allyn W. Rickett (Guan and Rickett 1985, 2001).

[39]In the chapter the *Annals of the Xia* in the *Records of the Grand Historian* (Shiji, ch. 2.1–2.33, pp. 51–67; translated by Allen 1895, pp. 93ff), an ancient historical record that is attributed to Sima Qian and published in the Former Han Dynasty (206 BCE–24 CE) around 100 BCE, as well as in

'unusual phenomena' (*zaiyi*) like floods and droughts, resulting in the decay of a country's waterways as expressed in the dynastic histories, are regarded as celestial warnings and are interpreted with regard to human affairs, particularly the emperor's rule that is not virtuous, benevolent and sincere.[40]

The idea of a celestially sanctioned imperial power had been transmitted throughout history, because it serves to show the rulers' power and strengthen the centralised government. With the Siberian Rivers Reversal model, Guo Kai draws on the strong connection between river control and governing a country, thus providing a good argument for the internal discourse in China. On other grounds, by contending that the Soviet Union's disintegration has been caused by mismanagement of its rivers and simultaneously pointing to the traditional knowledge of a connection between river control and state power, he claims that China not only can learn from failures of foreign countries but most of all should learn from her own historical knowledge. Drawing on Chinese traditional ideas in this context aims at strengthening the own discursive arguments for the Great Western Route. It further aims at raising a patriotic awareness that Chinese knowledge preceded Western knowledge with the intention of easing the perceived asymmetry between China's current status and the one she aspires to.

3.3 Conclusion

The discourse in favour of the Great Western Route appropriates foreign and traditional Chinese arguments. The respective models serve as signs of authority in the internal Chinese debate and thus are fashioned to meet its special requirements. The Great Western Route in this discourse is praised not only as a panacea against North China's desertification problem but also as a means of accomplishing diverse sociopolitical goals. The arguments are shaped to address the issues concerning the current government, such as the strengthening of China's economic power as

---

the chapter *The counsels of Great Yu* in the *Book of Documents* (Shangshu, ch. 3.2–3.23, pp. 80–90; translated by Legge 1960, pp. 52ff), a collection of ancient official speeches and documents of the first millennium BCE and of later date, a connection between Yu's engagement for the society in flood control and his achievement in state power is made. Because of his achievements, he had been appointed by the legendary ruler Shun as his descendant. Yu then founded the Xia Dynasty (twenty-first to sixteenth century BCE) and united the different tribal groups. He is therefore praised as the great engineer ruler.

[40]One example with frequent mention of such natural disasters is the reign of emperor Cheng (r. 33-7 BCE) of the Former Han Dynasty. For an example of abnormal natural phenomena that are brought in relation to human affairs, see the third year (30 BCE) of Emperor Cheng's reign in the dynastic history of the Former Han Dynasty (Hanshu 1, ch. 10, pp. 306–308). For a translation of the passage, see Homer H. Dubs' *History of the Former Han Dynasty* (Pan 1954, pp. 380ff). For some general remarks about the existence of abnormal natural phenomena and the interpretation of Confucian officials of these 'disasters' and 'unusual phenomena', see Dubs' introduction to the chapter of Emperor Cheng's reign (Pan 1954, pp. 363ff).

depicted with the TVA model, recovery of a respected position in the international arena and restoration of Chinese self-esteem as illustrated with the Rhine-Main-Danube Canal, and the assurance of national integrity and social stability as expressed through the Siberian Rivers Reversal Project. Rather than showing how China should learn from the 'West', the discourse participants translate foreign models of water diversion into the Chinese context by selectively appropriating certain arguments deemed useful within the internal Chinese debates. Moreover, they combine these foreign models with newly interpreted traditional Chinese knowledge in order to raise patriotism and national identity. Examples are the Grand Canal and traditional knowledge of effective river control as signs of benevolent rule. The former is depicted as a model that spread social development throughout the world earlier than Western knowledge, thereby turning the influx of Western knowledge in the early twentieth century into a kind of reimportation of original Chinese expertise. The latter concept preceded the Soviet Union's experience of disintegration that is assumed to have been caused by noneffective river control, thereby implying that the Soviet Union could have avoided its demise by applying traditional Chinese knowledge. By complementing foreign discourses with Chinese traditional knowledge that is depicted to have preceded Western knowledge, the advocators of the Great Western Route aim at easing the apparent asymmetry between Chinese self-perception of her cultural achievements and the humiliation felt in the face of 'modern' challenges. Thus, they intend to strengthen China's patriotism and sense of identity, as well as regain her assumed 'natural' position as centre of the world and as the harmonious and benevolent leader. The message is that China not only should learn from Western societies how to conduct her self-strengthening and modernisation attempt but most of all should learn from her own cultural heritage.

The arguments in favour of the Great Western Route all deal with recent problems that the central government is concerned about. North China's aridity is regarded as a critical obstacle for China's development attempt and a threat to China's sociopolitical stability and the country's integrity. Moreover, it is considered as impeding her ability to regain the aspired position of a spiritual centre of the world, which had been fundamentally damaged by the forced contact with the Western countries in the late imperial and republican era. In spite of official rhetoric of a 'harmonious coexistence of mankind with nature', the Chinese government's arguments are more based on ideas of dominating and transforming nature. Though promoting water saving measures, the *Comprehensive report on the strategic study on sustainable development of China's water resources* relies in its argumentation on the desirable effects of transformation and improvement of nature with large-scale hydro-engineering projects (Qian and Zhang 2001). Instead of managing the general planning of the ecosystem and the distribution of natural resources in compliance with the motto of 'define the needs according to the supply' (Fan 2007, p. 79), such large-scale water diversion projects 'define the supply according to the needs' (2007, p. 79), as the geologist Fan Xiao[41] in a publication of the environmental nongovernmental organisation Friends of Nature (*Ziran zhi you*) states.

According to the discourse analysed in this chapter, a water diversion project, greater than the official South-North Water Diversion Scheme, that fundamentally and on a long-term basis could solve North China's problem of aridity, such as the Great Western Route, therefore would not only correspond to the recent government's policy at work of constructing the most prestigious hydro-engineering projects worldwide that function as icons of national pride but would also attain her aspired goals of regaining a strong international position. Consequently it is not astonishing to find this topic back on the agenda after a period of relative silence. High-ranking officials of the water bureaucracy, who are also involved in feasibility studies of the Western Route, propose this route once again to solve North China's water shortage, which is claimed to become increasingly severe due to a combination of climate change, ground water depletion and seasonal droughts (see Zhang 2011). Whether it will be accepted by the Ministry of Water Resources remains to be seen. However, gigantic engineering tasks to tap Tibetan water resources seem to have an inspiring charm on Chinese hydro-engineers.

Glossary

Names and Dates

Deng Yingtao 邓英淘 (b. 1952)
Duan Ruofei 段若非 (b. 1936)
Fan Xiao 范晓
Guan Zhong 管仲 (eighth to seventh century BCE)
Gun 鲧, mythological figure
Guo Kai 郭开 (b. 1933)
Guo Shoujing 郭守敬 (1231–1316)
Hu Jintao 胡锦涛 (b. 1942)
Huang Wanli 黄万里 (1911–2001)
Jiang Zemin 江泽民 (b. 1926)
Li Peng 李鹏 (b. 1928)

[41] Fan Xiao, chief engineer at the Sichuan Bureau of Geological Exploration and Exploration of Mineral Resources, has become well known in Western media because of his concerns about hydropower dams in seismically active regions. He, among other Chinese and US scientists, suggested that the large earthquake in Wenchuan at the Min River in May 2008 had been induced by the weight of the close Zipingpu Dam's reservoir. A collection of articles concerning this topic can be found at the website *EastSouthWestNorth* under the headline 'Did the Zipingpu Dam Cause the Sichuan Earthquake?' (Soong 2009).

Li Yizhi 李仪祉 (1882–1938)
Lin Yishan 林一山 (1911–2007)
Mao Zedong 毛泽东 (1893–1976)
Qian Zhengying 钱正英 (b. 1923)
Shun 舜, mythological figure
Sima Qian 司马迁 (second to first century BCE)
Sun Yatsen (Sun Zhongshan) 孙中山, also known as Sun Wen 孙文 (1866–1925)
Wang Shucheng 汪恕诚 (b. 1941)
Wang Xiaoqiang 王小强 (b. 1952)
Wen Jiabao 温家宝 (b. 1942)
Weng Wenhao 翁文灏 (1889–1971)
Yao 尧, mythological figure
Yu 禹, mythological figure, also known as Great Yu
Yuan Jiazu 袁嘉祖 (b. 1930)
Zhang Zhengbin 张正斌 (b. 1962)

Subject Matters, Places, Titles, Institutions, Etc

Da xixian 大西线 (Great Western Route)
Guanzi 管子 (Guanzi; classical Chinese text)
Guoshi xuehui 国史学会 (Association of National History)
Minzu tuanjie 民族团结 (national/ethnic unity)
Nanshui beidiao gongcheng 南水北调工程 (South-North Water Diversion Project)
Ren yu ziran hexie xiangchu 人与自然和谐相处 (Harmonious coexistence of mankind with nature)
Shehui zhuyi hexie shehui 社会主义和谐社会 (Socialist harmonious society)
Shuo-Tian yunhe 朔天运河 (Shuo-Tian Canal)
Shuomatan 朔玛滩 (Shuomatan; proposed dam on the Yarlung Tsangpo)
Xibu da kaifa 西部大开发 (Great Western Development)
Xi xue Zhong yuan 西学中源 (Chinese origin of Western science)
Zaiyi 灾异 (disaster and unusual [natural] phenomena)
Zhongguo nongcun fazhan wenti yanjiuzu 中国农村发展问题研究组 (China Rural Development Research Group)
Zhongguo zhengce kexue yanjiuhui 中国政策科学研究会 (China Association of Policy Science)
Zhonghua minzu 中华民族 (Chinese people)
Ziran zhi you 自然之友 (Friends of Nature; Chinese environmental nongovernmental organisation)

References

Allen HL (1895) Ssŭma Ch'ien's historical records, art. III, chapter II. – the Hsia dynasty. J R Asiat Soc GB Irel (1895) 27(1):93–110

Arpi C (2008) Himalayan rivers: geopolitics and strategic perspectives. Indian Def Rev 23(2) (23 June 2008). http://www.indiandefencereview.com/2008/06/himalayan-rivers-geopolitics-and-strategic-perspectives.html. Accessed 9 Oct 2009

Birrell A (1993) Chinese mythology: an introduction. Johns Hopkins University Press, Baltimore

Cantian shuili ziyuan gongcheng yankaohui 参天水利资源工程研考会 [Study and Inspection Group of the Cantian Water Conservancy and Resources Project] (1999) Yuan Jiazu tan da xixian diaoshui fang'an 袁嘉祖谈大西线调水方案 [Yuan Jiazu talks about the Great Western Route Water Diversion Plan]. Cantian shuili ziyuan gongcheng yankaohui "Gongzuo tongbao" 参天水利资源工程研考会《工作通报》 ["Working Circular" of the Study and Inspection Group of the Cantian Water Conservancy and Resources Project] 8 (25 February 1999). http://www.ecrcass.com/ecrcass/CTTB/1999/1999_8.htm. Accessed 2 Feb 2009

Cantian shuili ziyuan gongcheng yankaohui 参天水利资源工程研考会 [Study and Inspection Group of the Cantian Water Conservancy and Resources Project] (2000) Lin Yishan tan da xixian diaoshui yu Xibu da kaifa 林一山谈大西线调水与西部大开发 [Lin Yishan talks about the Great Western Route Water Diversion and the Great Opening-up of the West]. Cantian shuili ziyuan gongcheng yankaohui "Gongzuo tongbao" 参天水利资源工程研考会《工作通报》 ["Working Circular" of the Study and Inspection Group of the Cantian Water Conservancy and Resources Project] 2 (8 September 2000). http://www.ecrcass.com/ECRCASS/CTTB/2000/2000_zk2.htm. Accessed 18 Dec 2009

Chi Ch-T (1963) Key economic areas in Chinese history, as revealed in the development of public works for water-control, 1st edn., 1936. Paragon Book Reprint Corporation, New York

Davydov MM (1972) The Ob will enter the Caspian: the Yenisey-Ob-Aral-Caspian water connection and the energy problem. Sov Geogr Rev Transl 13(9):603–617. Translated from Sibirskiye ogni [Siberian Fires] (1949) 2:102–110

Deng, Yingtao 邓英淘, Wang, Xiaoqiang 王小强, Cui, Jiming 崔鹤鸣, Yang, Shuang 杨双 (1999) Zai zao Zhongguo 再造中国 [Rebuild China]. Wenhui chubanshe, Shanghai

Duke DF (2006) Seizing favours from nature: the rise and fall of Siberian river diversion. In: Tvedt T, Jakobsson E (eds) A history of water. Water control and river biographies. I. B. Tauris, London, pp 3–34

Eckholm E (1999) Detour on capitalist road: die-hard Maoist collective. The New York Times (7 January 1999), A1

Ekbladh D (2001) "Mr. TVA": grass-roots development, David Lilienthal, and the rise and fall of the Tennessee Valley Authority as a symbol for U.S. Overseas Development, 1933–1973. Dipl Hist 26(3):335–374

Elvin M (2004) The retreat of the elephants: an environmental history of China. Yale University Press, New Haven

Environment and Development Desk (2007) Damming Tibetan waters. In: Environment and Development Desk, Department of Information and International Relations, Central Tibetan Administration (ed) A human development and environment report, Tibet, pp 121–148. http://www.tibet.net/en/index.php?id=95&rmenuid=11. Accessed 6 Nov 2009

Fan, Xiao 范晓 (2007) Daxing gongcheng dui huanjing de yingxiang 大型工程对环境的影响 [The environmental impacts of large-scale construction projects]. In: Yang Dongping 杨东平 (ed) 2006 nian: Zhongguo huanjing de zhuanxing yu boyi 2006年:中国环境的转型与博弈 [2006: changes and struggles of China's environment], Huanjing lüpishu 环境绿皮书 [Environment yearbook]. Shehui kexue wenxian chubanshe 社会科学文献出版社, Beijing, pp 70–79 of pdf version. http://www.fon.org.cn/upload/download/200711/20071129112720_2006fullversion-FONgreenpaper.pdf. Accessed 23 Oct 2009

Fan X (2008) The environmental impacts of large-scale construction projects. In: Yang DP (ed) The China environment yearbook, Vol 2: Changes and struggle. Brill, Leiden, pp 189–213

Feng C (1999) An unfinished battle in China: the leftist criticism of the reform and the third thought emancipation (research note). China Q 158:447–467
Fewsmith J (1994) Dilemmas of reform in China: political conflict and economic debate (Socialism and Social Movements). Sharpe, Armonk
Giese E (1998) Die ökologische Krise des Aralsees und der Aralseeregion: Ursachen, Auswirkungen, Lösungsansätze. In: Giese E, Bahro G, Betke D (eds) Umweltzerstörungen in Trockengebieten Zentralasiens (West- und Ost-Turkestan): Ursachen, Auswirkungen, Maßnahmen (Erdkundliches Wissen 125). Steiner, Stuttgart, pp 55–119
Goodman DSG (2004) The campaign to "Open up the West": national, provincial-level and local perspectives. China Q 178:317–334
Guan Zh, Rickett AW (1985) Guanzi: political, economic, and philosophical essays from early China: a study and translation, vol 2. (trans: Rickett AW) (Princeton Library of Asian Translations). Princeton University Press, Princeton
Guan Zh, Rickett AW (2001) Guanzi: political, economic, and philosophical essays from early China: a study and translation, vol 1. (trans: Rickett AW) (Princeton Library of Asian Translations). Princeton University Press, Princeton
Guanzi 管子, compiled by Guan, Zhong 管仲 and Fang, Xuanling 房玄龄. (Qianlong yulan Siku quanshu huiyao 乾隆御览四库全书荟要 58). Jilin renmin chubanshe, Changchun, pp 1–284, 1997
Guo, Kai 郭开 (1998) Guanyu da xixian diaoshui gongcheng 关于大西线调水工程 [About the Great Western Route Water Diversion Project]. Neibu canyue 内部参阅 [Internal Consultative Readings] 12:2–9
Guo, Kai 郭开 (2001) Diaoshui zhisha, xingwang suo xi 调水治沙 兴亡所系 [Water transfer and desert control have an impact on [China's] rise and fall]. Dangdai sichao 当代思潮 [Trends in Contemporary Thoughts] 6:50–56
Hanshu 汉书 [The books of the [former] Han [dynasty]], compiled by Ban, Gu 班固 and Yan, Shigu 颜师古. Zhonghua shuju, Beijing, 1975
Holbig H (2004) The emergence of the campaign to open up the West: ideological formation, central decision-making and the role of the provinces. China Q 178:335–357
Kao MYM, Leung JK (eds) (1986) The writings of Mao Zedong, vol 1. Sharpe, Armonk
Legge J (1960) The books of Yu. Book II: the counsels of the Great Yu. In: Legge J (ed) The Shoo king or the Book of Historical Documents (The Chinese Classics 3). Hong Kong University Press, Hong Kong, pp 52–67
Leonard JK (1996) Controlling from afar: the Daoguang emperor's management of the Grand Canal crisis, 1824–1826 (Michigan monographs in Chinese studies 69). Center for Chinese Studies, University of Michigan, Ann Arbor
Lewis ME (2006) The flood myths of early China (SUNY Series in Chinese Philosophy and Culture). State University of New York Press, Albany
Li, Ling 李伶 (2006) Xizang zhi shui jiu Zhongguo: Da xixian "zai zao Zhongguo" zhanlüe neimu xianglu 西藏之水救中国:大西线"再造中国"战略内幕详录 [Save China through water from Tibet: a detailed record of the inside story of the Great Western Route "Rebuild China" strategy], 2nd edn. Zhongguo Chang'an chubanshe, Beijing
Li C, White L (1990) Elite transformation and modern change in mainland China and Taiwan: empirical data and the theory of technocracy. China Q 121:1–35
Lilienthal DE (1944) TVA: democracy on the march, 8th edn. Harper, New York
Liu, Hong 柳红 (2009) Jiaota shidi de Nongfazu 脚踏实地的农发组 [The Earnest and Down-to-earth Rural Development Group]. Zhongguo xuanju yu zhili 中国选举与治理 [China Elections and Governance] (14 December 2009). http://www.chinaelections.com/printnews.asp?newsid=163574. Accessed 17 Feb 2010. Original source: Jingcha guanchabao 经济观察报 [Economic Observer News]
MacFarquhar R (1983) The Great Leap Forward, 1958–1960 (Origins of the Cultural Revolution 2). Oxford University Press, Oxford
Mao, Ronglin 茅荣林 (1947) YVA yu Zhongguo minzhu Y•V•A 与中国民主 [YVA and Chinese democracy]. Zhongyang ribao 中央日报 [Central Daily], (Minguo 36.04.17-19/5-7 June 1947), 5

McCauley M (1976) Khrushchev and the development of Soviet agriculture: the virgin land programme 1953–1964 (Studies in Russian and East European History). Macmillan, London

N. N., Guo, Kai 郭开 (1998a) Nanshui beidiao de hongwei gouxiang – fang shuili zhuanjia Guo Kai 南水北调的宏伟构想——访水利专家郭开 [The grand concept of the South-North Water Diversion Scheme – interview with the water resources specialist Guo Kai]. Dangdai sichao 当代思潮 [Trends in Contemporary Thoughts] 1:9–17

N. N., Guo, Kai 郭开 (1998b) Nanshui beidiao de hongwei gouxiang – zai fang shuili zhuanjia Guo Kai 南水北调的宏伟构想——再访水利专家郭开 [The grand concept of the South-North Water Diversion Scheme – a second interview with the water resources specialist Guo Kai]. Dangdai sichao 当代思潮 [Trends in Contemporary Thoughts] 3:2–9

N. N., Guo, Kai 郭开 (1998c) Nanshui beidiao de hongwei gouxiang – sanfang shuili zhuanjia Guo Kai 南水北调的宏伟构想——三访水利专家郭开 [The grand concept of the South-North Water Diversion Scheme – a third interview with the water resources specialist Guo Kai]. Dangdai sichao 当代思潮 [Trends in Contemporary Thoughts] 5:2–13

N. N., Guo, Kai 郭开 (2006) Xizang zhi shui neng fou jiu ganhan Zhongguo? 西藏之水能否救干旱中国? [Can Tibetan water save arid China?]. In: Xizang zhi shui qiu Zhongguo 西藏之水求中国 [Save China through water from Tibet], Zhongguo Xizang xinxi zhongxin 中国西藏信息中心 [China Tibet Information Center] (10 October 2006). http://info.tibet.cn/zt2006/xzzsjzg/qtsp/200610/t20061024_167174.htm. Accessed 16 Feb 2010. Original source: Beijing kejibao 北京科技报 [Beijing Science and Technology News]

Pan K (1954) First division: the imperial annals, chapters VI-X, 2. In: Dubs HH (ed) The history of the former Han dynasty (trans: Dubs HH). Waverly Press, Baltimore

Pietz D, Giordano M (2009) Managing the Yellow River: continuity and change. In: Molle F, Wester P (eds) River basin trajectories: societies, environments and development (Comprehensive assessment of water management in agriculture series 8). CAB International, Oxfordshire, pp 99–122

Pomeranz KL (2009) The great Himalayan watershed: agrarian crisis, mega-dams and the environment. New Left Rev 58:5–39

Qian, Zhengying 钱正英 (2001) Zhongguo shui ziyuan zhanlüe yanjiu zhong jige wenti de renshi 中国水资源战略研究中几个问题的认识 [About some problems in strategic research of China's water resources]. Hehai daxue xuebao (Ziran kexueban) 河海大学学报(自然科学版) [Journal of Hehai University (Natural Sciences)] 3:1–7

Qian, Zhengying 钱正英, Zhang, Guangdou 张光斗 (eds) (2001) Zhongguo kechixu fazhan shuiziyuan zhanlüe yanjiu zonghe baogao ji ge zhuanti baogao 中国可持续发展水资源战略研究综合报告及各专题报告 [Comprehensive report on the strategic study on sustainable development of China's water resources and different position papers], Zhongguo gongchengyuan zhongda zixun xiangmu: Zhongguo ke chixu fazhan shuiziyuan zhanlüe yanjiu zonghe baogaoji 中国工程院重大咨询项目:中国可持续发展水资源战略研究综合报告集 [Great project in consultation with the Chinese Academy of Engineers: collection of the comprehensive report on the strategic study on sustainable development of China's water resources] 1. Zhongguo shuili shuidian chubanshe, Beijing

RMRB (Renmin ribao) (2003) Zhongguo 21 shiji chu kechixu fazhan xingdong gangyao 中国21世纪初可持续发展行动纲要 [China's sustainable development agenda in the early 21st century]. Renmin ribao 人民日报 [People's Daily] (25 July 2003), 6

Russel D (1949) The TVA idea. The Foundation for Economic Education, Irvington-on-Hudson

Schoenhals M (1987) Saltationist socialism: Mao Zedong and the Great Leap Forward 1958 (Skrifter utgivna av Föreningen för Orientaliska Studier 19). Department of Oriental Languages, University of Stockholm, Stockholm

Seeger M (2013) Water diversion between official rhetoric and the drive to dominate nature. In: Meinert C (ed) Nature, environment, and culture in East Asia: the challenge of climate change, Climate and culture 1. Brill, Leiden, pp 163–195

Shangshu zhushu 尚书注疏 [Annotated edition of the book of documents], compiled by Kong, Anguo 孔安国, Lu, Deming 陆德明 and Kong, Yingda 孔颖达 (Qianlong yulan Siku quanshu huiyao 乾隆御览四库全书荟要 4). Jilin renmin chubanshe, Changchun, 1a-472b, 1997

Shapiro J (2002) Mao's war against nature: politics and the environment in revolutionary China, 2nd edn, Studies in environment and history. Cambridge University Press, Cambridge

Shenbao (1938) Dao Li Yizhi xiansheng 悼李仪祉先生 [In mourning for Mr. Li Yizhi]. Shenbao 申报 [The Shun Pao] (Minguo 27.03.10/10 April 1938), 1

Shiji 史记 [Records of the Grand Historian], compiled by Sima, Qian 司马迁, Fei, Yin 裴骃, Sima, Zhen 司马贞 and Zhang, Shoujie 张守节 (Qianlong yulan Siku quanshu huiyao 乾隆御览四库全书荟要 26). Jilin renmin chubanshe, Changchun, 1997

Shiva V (2006) Erd-Demokratie: Alternativen zur neoliberalen Globalisierung. Rotpunktverlag, Zürich

Shui ziyuan diaopei yu guotu zhengzhi ketizu 水资源调配与国土整治课题组 [Study Group of Allocation of Water Resources and Renovation of the National Territory] (2000a) Da xixian diaoshui yu Xibu dakaifa 大西线调水与西部大开发 [About the Great Western Route Water Diversion and the Great Opening-up of the West]. Cantian shuili ziyuan gongcheng yankaohui "Gongzuo tongbao" 参天水利资源工程研考会《工作通报》 ["Working Circular" of the Study and Inspection Group of the Cantian Water Conservancy and Resources Project] 8 (12 April 2000). http://www.ecrcass.com/ecrcass/CTTB/2000/2000_8.htm. Accessed 18 Dec 2009

Shui ziyuan diaopei yu guotu zhengzhi ketizu 水资源调配与国土整治课题组 [Study Group of Allocation of Water Resources and Renovation of the National Territory] (2000b) Zai tan da xixian diaoshui yu Xibu da kaifa 再谈大西线调水与西部大开发 [Once again about the Great Western Route Water Diversion and the Great Opening-up of the West]. Cantian shuili ziyuan gongcheng yankaohui "Gongzuo tongbao" 参天水利资源工程研考会《工作通报》 ["Working Circular" of the Study and Inspection Group of the Cantian Water Conservancy and Resources Project] 1 (6 September 2000). http://www.ecrcass.com/ecrcass/CTTB/2000/2000_zk1.htm. Accessed 18 Dec 2009

Shuili ziyuan diaopei yu guotu zhengzhi ketizu 水利资源调配与国土整治课题组 [Study Group of Allocation of Water Resources and Renovation of the National Territory] (1999) Zai zao yige Zhongguo: shuili ziyuan diaopei yu guotu zhengzhi de jige wenti 再造一个中国——水利资源调配与国土整治的几个问题 [Rebuild another China: some questions about the allocation of water resources and the renovation of the national territory]. Zhongguo shehui kexueyuan jingji wenhua yanjiu zhongxin – diaocha yanjiu tongxun 中国社会科学院经济文化研究中心——调查研究通讯 [Chinese Academy of Social Sciences' Economy and Culture Research Center: Survey and Research Newsletter] 2 (3 October 1999). http://www.ecrcass.com/ECRCASS/DYTX/1999/1999_2.htm. Accessed 12 Feb 2010

Smil V (2009) Finding mutual interests in nature. Far East Econ Rev 172(8):44–47

Soong R (2009) Did the Zipingpu Dam cause the Sichuan earthquake? Dong nan xi bei 东南西北 [EastSouthWestNorth] (7 February 2009). http://www.zonaeuropa.com/20090207_1.htm. Accessed 6 Nov 2009

Sun YS (1929) The international development of China, 2nd edn. Hutchinson, London

Tian XW (2001) Deng Xiaoping's Nanxun: impact on China's regional development. In: Wong J, Zheng YN (eds) The Nanxun legacy and China's development in the post-Deng era. Singapore University Press/World Scientific Publishing, Singapore, pp 75–92

Tsering T (2002) HydroLogic: water for human development. An analysis of China's water management and politics. In: Tibet Justice Center (ed) Tibet: current environment and development issues. http://www.tibetjustice.org/reports/hydrologic.pdf. Accessed 17 Nov 2009

Veblen T (1921) The engineers and the price system. Huebsch, New York

Wang, Huayun 王化云 (1957) Mao zhuxi shicha Huanghe ji 毛主席视察黄河记 [Record of chairman Mao's inspection of the Yellow River]. Jiefang ribao 解放日报 [Liberation Daily] (6 April 1957), 1 and 3

Wang, Weiqun 王伟群 (1999) Zai zao yige Zhongguo 再造一个中国 [Rebuild another China]. Zhongguo qingnianbao 中国青年报 [China Youth Daily] (19 July 1999). http://daxixian.blog.hexun.com/8237578_d.html. Accessed 28 Feb 2010. Original source: http://www.cyol.net/cyd/zqb/19990719/GB/9564%5EQ511.htm

Wang, Shucheng 汪恕诚 (2004) Zai tan ren yu ziran hexie xiangchu – jianlun daba yu shengtai 再谈人与自然和谐相处——兼论大坝与生态 [Speaking again about the harmonious coexistence of mankind with nature – simultaneously speaking about dams and ecology]. Zhongguo shuili 中国水利 [China Water Resources] 8:6–14

Wang, Xiaoqiang 王小强 (2009) Chongti xibu da kaifa – qidong neixu yu chanye shengji de fanglüe 重提西部大开发——启动内需与产业升级的方略 [Raising again [the topic of] the Great Western Development – the strategy of starting internal demand and industrial upgrading]. Lüye 绿叶 [Green Leaf] 6:92–96

Wang, Mingyun 王明运, Ye, Qingchao 叶青超, Ling, Meihua 凌美华 (1992) Shuo-Tian yunhe wenji 朔天运河文集 [Collected texts of the Shuo-Tian canal]. Qixiang chubanshe 气象出版社, Beijing

WCED (Commission on Environment and Development, United Nations) (1987) Report of the World Commission on Environment and Development: our common future. http://www.un-documents.net/wced-ocf.htm. Accessed 3 Nov 2009

Wirth E (1983) Die Ohnmacht der Vernunft: Vom Sinn und Unsinn eines Schiffahrtsweges Rhein-Main-Donau. In: Weiger H (ed) Der Rhein-Main-Donau-Kanal: Das Für und Wider seiner Fertigstellung (Iris-Bücher Dokumentation 504). Klaus Schulz Verlag, Nürnberg, pp 45–81

Wong GHC (1963) China's opposition to Western science during late Ming and early Ch'ing. Isis 54(1):29–49

Yuan, Jiazu 袁嘉祖 (1998) Nanshui beidiao, ke bu ronghuan 南水北调 刻不容缓 [The South-North Water Diversion Scheme, wasting no time]. Dangdai sichao 当代思潮 [Trends in Contemporary Thoughts] 4:2–10

Zhang K (2011) Diversion debate. China Dialogue (13 June 2011). http://www.chinadialogue.net/article/show/single/en/4349-Diversion-debate. Accessed 14 June 2011

Zhang, Zhengbin 张正斌, Duan, Ziyuan 段子渊, Xu, Ping 徐萍 (2008) Jiejue Zhongguo ganhan queshui wenti he fazhan quyu xiandai nongye fanglüe tantao 解决中国干旱缺水问题和发展区域现代农业方略探讨 [Strategic discussion for solving China's problem of aridity and lack of water and developing a zone of modern agriculture]. Zhongguo kexueyuan 中国科学院 [Chinese Academy of Sciences] (4 January 2008). http://daxixian.blog.hexun.com/16976623_d.html. Accessed 6 Feb 2010. Original source: http://www.cas.cn/xw/zjsd/200906/t20090608_646999.shtml

Zhao, Cheng 赵诚 (2001) Dan jiao mo jue Guanglingsan – ji Zhongguo shuili zhuanjia Huang Wanli xiansheng 但教莫绝广陵散——记中国水利专家黄万里先生 ["If only [the transmission of my] teachings will not cease [like the music piece] Guanglingsan [by Ji Kang 嵇康 of the state of Wei during the Three Kingdoms]" – in memorial to the Chinese water conservancy expert Mr. Huang Wanli]. In: "Huang Wanli wenji" bianji xiaozu 《黄万里文集》编辑小组 ["Collected Works of Huang Wanli" Editorial Group] (ed) Huang Wanli wenji 黄万里文集 [Collected Works of Huang Wanli]. Qinghua University, Beijing, pp 340–347. http://www.sciencenet.cn/upload/blog/file/2010/3/201032111539859865.pdf. Accessed 20 Jan 2011

Zhongguo Gongchandang di shiliu jie Zhongyang weiyuanhui di si ci quanti huiyi 中国共产党第十六届中央委员会第四次全体会 [Fourth plenum of the 16th Chinese Communist Party Central Committee] (2004) Zhonggong Zhongyang guanyu jiaqiang dang de zhizheng nengli jianshe de jueding 中共中央关于加强党的执政能力建设的决定 [Resolution of the Chinese Communist Party Central Committee on strengthening the party's governmentality] (19 September 2004). http://www.china.com.cn/chinese/2004/Sep/668376.htm. Accessed 2 Feb 2010

Zhongguo Gongchandang di shiliu jie Zhongyang weiyuanhui di liu ci quanti huiyi 中国共产党第十六届中央委员会第六次全体会议 [Sixth plenum of the 16th Chinese Communist Party Central Committee] (2006) Zhonggong Zhongyang guanyu goujian shehui zhuyi hexie shehui ruogan zhongda wenti de jueding 中共中央关于构建社会主义和谐社会若干重大问题的决定 [Resolution of the Chinese Communist Party Central Committee on some major issues regarding the building of a harmonious socialist society]. Zhongguo zhengfu ziliaoku 中国政府资料库 [Database of the Chinese Government]

Zhongguo shehui kexueyuan jingji wenhua yanjiu zhongxin 中国社会科学院经济文化研究中心 [Economy and Culture Research Center at the Chinese Academy of Social Sciences] (n.d.) Zhongguo jingji wenhua yanjiu 中国经济文化研究 [China economy and culture research]. http://www.ecrcass.com/ECRCASS/index1.htm. Accessed 18 Dec 2009

Zhongguo shehui kexueyuan jingji wenhua yanjiu zhongxin "Guoji wenti" yanjiuzu 中国社会科学院经济文化研究中心《国际问题》研究组 ["International Problems" Study Group of the Centre for Economic and Culture Research at the Chinese Academy of Social Sciences] (1998) Zai zao yige Zhongguo – Shuotian yunhe jianjie ji qi yiyi fenxi 再造一个中国——朔天运河简介及其意义浅析 [Rebuild another China – a short introduction of the Shuo-Tian Canal and a sharp analysis of its significance]. Dangdai sichao 当代思潮 [Trends in Contemporary Thoughts] 6:34–44

Zhonghua renmin gongheguo Guowuyuan xinwen bangongshe 中华人民共和国国务院新闻办公室 [Information Office of the State Council of the People's Republic of China] (n.d.) Zhongguo zhengfu baipishu 中国政府白皮书 [White papers of the Chinese government]. http://www.china.com.cn/ch-book/index.htm (Chinese), http://china.org.cn/e-white/index.htm (English). Accessed 6 Jan 2010

Chapter 4
Filling Multipurpose Reservoirs with Politics: Displacing the Modern Large Dam in India

Rohan D'Souza

It is the Multi-purpose storage reservoir that has correctly come to symbolize the engineering of the era of nation-building

(Hart 1956, p. 256).

Politically, they [large dams] have become a weapon for the rich, urban and powerful to take control of water resources away from the poor, rural and dispossessed

(Pearce 2005, p. 99).

Abstract Large dams have often been declared as purely expert-driven techno-economic artefacts which are beyond politics. This is frequently reflected in the calculation of cost-benefit ratios, which are assumed by many to objectively assess a dam's social, economic and ecological impacts. This assumption is increasingly disputed by studies which instead suggest that outcomes are often arrived at politically and directed towards realising specific political goals.

Keywords Large dam • Supply-side hydrology • Cost-benefit ratio • Political reasoning • India

As a rookie civil engineer in the early 1950s, Dr. K. L. Rao (later to become Union Minister for Irrigation and Power (1963–1973) in the Central Government) was deputed to scout around Andhra Pradesh (South India) to site the Nagarjuna Sagar dam. But the task of carrying out this most basic survey was soon complicated when he found himself and his engineering staff in the middle of an active, communist-led

R. D'Souza (✉)
Centre for Studies in Science Policy, School of Social Sciences,
Jawaharlal Nehru University, New Delhi 110 067, India
e-mail: rohanxdsouza@gmail.com

M. Nüsser (ed.), *Large Dams in Asia*, Advances in Asian Human-Environmental Research, DOI 10.1007/978-94-007-2798-4_4,
© Springer Science+Business Media Dordrecht 2014

guerilla insurgency. For Dr. Rao, preventing what he presumed was an apolitical-technical mission from being dragged into the messy struggle against landlordism in the survey zone, could not have been more perplexing. In time, however, he was to realise, as he notes in his memoirs, that plans for large dams tend to evoke an altogether unexpected response.

> Later, I got a letter from the leader of the Communist Party who was underground, that there was no necessity for me to have a police escort and that they would not have harmed me and the other engineers unless we were engaged in building roads to their hide-outs. Engineers dealing with dams and irrigation projects were most welcome. This was similar to what the Communists told Dr. Savage when he went to the river Yangtze in China to see a storage dam site. The Communists sent word to him that he could freely move about without escort as they would not harm engineers engaged in the development of rivers (Rao 1978, p. 37).

In fact, for several decades following 1947, the large dam in India was often presented as a political conundrum and voiced in contradictory claims. Jawaharlal Nehru, India's first Prime Minister (1947–1964), offered one such tone in an often quoted speech made in July 1954 in which he likened the large dam to a 'modern temple'. Subsequently, in a less remembered speech made before a gathering of engineers and technocrats in 1958, Nehru, as if in contrition, bemoaned the quest for big dams as a 'disease of gigantism' (Roy 1999, p. 104).[1] Nehru's contrary views were, perhaps, understandable for the times. The post-Second World War denouement was unprecedented in several ways. A period that left unquestioned the idea of progress insisted upon the supreme belief in development, inculcated faith in modern technology and advocated an unwavering confidence in positivist science.[2]

To many, the large dam appeared as an object that was class neutral, if not beyond politics. Such was its seeming apolitical allure that Henry C. Hart, an American academic and commentator on India, declared with much gusto in a book published in the late 1950s that the large dam with its multipurpose reservoir had 'correctly come to symbolise the engineering of the era of nation-building' (Hart 1956, p. 256). Put differently, the 'development of rivers' seemed to have charged decolonising nations with a new technological mission: the giant quest to transform fluvial powers into national assets – hydroelectricity, navigation, irrigation and flood control.

Turning dammed rivers into synonyms for nation-building, however, did not spring unadulterated from the breasts of technology enthusiasts or their severe will to dominate flows. Rather, calculations for the modern large dam had been derived from many of the troubled forces that had overwhelmed capitalism in the early decades of the twentieth century – the Great Depression in the United States, the

---

[1]Gigantism as the condition of the modern positivist technocratic mind is well summed up by Paul Virilio's stark quip that these 'enthusiasts for Progress' are but a 'dangerous gang of dwarves smitten with gigantism', who in entertaining a naïve conception of the world have the 'satisfaction of a stubbornly repeated infantile refusal' (Virilio 2002, p. 2). Put differently, plagued by the refusal to grow up.

[2]For an excellent discussion on the emergence of the development mindset and its embrace of science and technology amongst Indian politicians and planners between the 1920s and 1940s, see Zachariah (2005).

crisis of capitalist over production and the brutal failings of the free market. It was in the resultant vortex of near desperate interventions to save capitalism through the New Deal, the Fordist compromise, Keynesian economic pump priming and the crafting of capitalist planning that the comprehensive control of the Tennessee river through a series of multipurpose large dams was assembled. Under the aegis of the Tennessee Valley Authority (TVA), dams over the Tennessee River were expected to transform the region into an economically dynamic and modern productive landscape (D'Souza 2003b). The TVA model was soon to mark a profound hydraulic departure by kick-starting the post Second World War global obsession for large dams.[3]

With a formidable collection of technocrats, water bureaucrats, engineers, sundry social experts and several charismatic chairmen, the TVA enthusiasts not only set about attempting total river control but also simultaneously aimed at obscuring the large dam's political roots. In other words, the large dam under the rubric of multipurpose river valley development (MPRVD) was declared a technology exorcised of politics: a pure expert-driven techno-economic artefact intended to dominate nature for the freedom of man, to pursue national triumph by disciplining rivers and to create abundance through controlled flows (see Scott 2006).

In many ways, India's colonial legacy made possible and reinforced the pursuit of hydraulic capitalism through the large dam. Historically, technologies for hydraulic manipulation in the Indian subcontinent have moved through three distinct though overlapping phases. From the earliest times, tanks, inundation canals, temporary structures to trap drainage, wells and waterwheels made up the ensemble of water-harvesting structures. These techniques were essentially directed towards either impounding precipitation, tapping river inundations or retrieving groundwater recharge.[4] At the risk of oversimplification, one could perhaps conclude that the underlying hydraulic principle was to adapt water-harvesting structures and designs to microclimates, topography and fluvial process.

In the early nineteenth century, however, British colonialism initiated a radical break in both technique and hydraulic principle by introducing perennial canal irrigation in several parts of the South Asian subcontinent. For the first time, permanent headworks in the form of barrages and weirs were thrown across riverbeds and their waters diverted through intricate and extensive canal systems. These barrages and weirs were equipped with a series of shutters to regulate flows by impounding water during lean seasons and diverting it into canals; and, on the reverse, the shutters could be flipped open to release waters during the river's peak discharge. In effect, by flattening the river's variable flow regime at certain points along its course, irrigation was transformed from a seasonal to a perennial

[3]On the global influence of the TVA and India in particular, see Klingensmith (2007). Also see Biggs (2006) and Hoag (2006).

[4]For a comprehensive discussion of the various types of water-harvesting structures termed 'traditional', see Agrawal and Narain (1997).

possibility. This phase, often referred to as the advent of the era of modern irrigation, witnessed the construction of several large canal irrigation schemes with permanent headworks such as the Ganges Canal (1854), the Godavari (1852) and the Krishna (1855) systems. These big-engineering efforts, in several ways, had profoundly transformative impacts. The civil engineer and the bureaucratic control of water, in particular, soon caused the expropriation of the skills of the local irrigator and unsettled the 'fluvial wisdom' of the community (D'Souza 2006a).

Changes in irrigation technologies were also followed up with dramatic alterations to entire hydraulic environments, the case in point being that of the Eastern deltas (contemporary Bengal, Bihar and Orissa), which were transformed from being flood-dependent agrarian regimes into flood-vulnerable landscapes (see D'Souza 2006b). In the quest for comprehensive flood control, the colonial dispensation undertook the systematic construction of flood control embankments to hem in rivers within their main channels. Though driven chiefly by the need to secure private property in land, these flood control measures soon disrupted natural flow regimes and ended up aggravating flood lines and thereby opening up the deltas to augmented flood vulnerability (Singh 2008). In addition, they also constructed a network of roads, railway lines and bridges, which by running in the east–west direction ended up interrupting natural drainage lines that mostly dropped from north to south. These structures, in time, not unexpectedly, began to unsettle a complex and fragile arrangement for drainage. By the beginning of the twentieth century, natural drainage arrangements survived only in pockets, as vast parts of eastern India had been transformed into a 'succession of water logged morasses' in which 'dismal swamps-breeding malaria', were debilitating the population and the fertility of the soil (see Whitcombe 1995).

Colonial hydraulic interventions, as it is now widely recognised, oversaw the dismantling and destruction of several unique water traditions in India. This, of course, is not to argue that all 'traditional' water practices were ideal, enduring and environmentally sound. Rather, the emphasis in this section is to point out that the contemporary model for harnessing water in India amplifies its colonial legacy by continuing to expropriate or eliminate traditional water management skills and technologies. And having thereby relentlessly extinguished other ways, techniques, arrangements, traditions and cultures for managing and conserving water in India, the large dam, spurred by supply-side hydrology, is always pursued as the TINA ('There Is No Alternative') option.

According to a recent count, over 45,000 large dams currently sit astride innumerable river valleys, gorges and 'gunshot' sites (WCD 2000). Formerly wild cascading flows are now put to work – running turbines, marching as orderly cusecs in irrigation canals, providing the measured electric hum for industrial machines, winding their way diligently through drinking water pipes or simply held as silent volumes in immense reservoirs. The river has been put on tap. A dammed river, however, as I will suggest below, most profoundly sets off a range of irreconcilable political tensions and ecological frictions. Modern large dams, given the experiences in the last 60 years in particular, have been deeply implicated in various processes integral to crafting social and economic inequalities such as enabling enclosure, transferring hydraulic endowments to powerful constituencies, intensifying monocropping and industrial agriculture, shifting the ecological costs onto marginal communities and being complicit in the expropriation and elimination of indigenous water management traditions.

4.1 Enclosing the River: When Dams Ate People

In India, disquiet over large dams was first expressed over the issue of displacement. The multipurpose reservoir, in requiring to create an artificial lake, had perforce to drown vast swathes of existing forests and habitations. Thus, entire villages and settled communities, which fell within the bed of the dam's reservoir, were forcibly evacuated from their lands and homes. By the 1980s, the number of 'oustees' or project-affected persons (PAPs) had reached such alarming proportions that the much celebrated *Second Citizens' Report* (Agarwal and Narain 1985), brought out to highlight the region's growing environmental crises, was dedicated 'to the dam-displaced people of India'.

Dam-displacement victims – 'refugees of an unacknowledged war' in the words of Arundhati Roy (1999, p. 24) – were doubly dispossessed. On the one hand, all their possible livelihood means were comprehensively destroyed through submergence, while, on the other, they were systematically denied any meaningful resettlement or rehabilitation (R&R). Initially, under the pretence of compensation, the PAPs were simply paid paltry cash settlements. In 1984, however, partly following the intense resistance that was building up against the infamous Sardar Sarovar Project (SSP) over the Narmada River, the official policy on R&R finally was to concede the right to a land for land compensatory package. Despite this seemingly radical possibility, the R&R strategy in India continues to act as a type of new enclosure. First, armed by the archaic colonial Land Acquisition Act of 1894, the government exercises *eminent domain* over all land, which can then be seized for anything that is deemed as a 'public purpose' requirement. Through such a legal framing, moreover, the dispossessed are also denied any right to either challenge or dispute the government's definition of what constitutes a public purpose. With their livelihoods thus lost, the oustees are then further compromised. The implementation of R&R programmes has invariably tended to address compensation claims by breaking whole communities that previously existed as culturally dense intertwined arrangements into now oversimplified family units. In effect, the deep associations that sustained and secured the viability of various kinds of social groupings (especially that of tribal or *Adivasis* communities) are disoriented and rendered instead, by design, into collections of atomised individuals. In other words, the bureaucratic and formal categories deployed to facilitate the economic calculations for R&R have led to the forced snapping of deep historical ties, bonds and cultural linkages that were critical to survival strategies and livelihood means. And lastly, by concentrating all efforts on estimating equivalences to land loss, the R&R strategy has ended up ignoring and devaluing an entire range of other subsistence institutions and means such as commonly shared forests, grasslands, streams, tanks, fishing rights and village commons, a web of natural endowments that were most depended upon by the landless, the marginal and the impoverished.[5]

Clearly, dam displacement in terms of both the legality of its direct seizure of livelihood means and in the details of enforced atomisation and increased individualised vulnerability amounts to a contemporary version of enclosure. One conservative estimate of the number of people displaced by large dams in India since 1947 is placed at 40 million, with possibly a mere tiny fraction of this huge number of oustees having managed anywhere near meaningful resettlement.

4.2 Political Arithmetic for Legitimacy

The large dam is always announced as a techno-economic decision. Typically, quantification is pursued, which, in the main, boils down to the search for an acceptable cost-benefit (CB) ratio for the project. The CB ratio is aimed at providing a measure of the return on investment for every unit of capital invested over the lifecycle of the project. Calculating the CB involves attempts at neatly lining up, like a chartered accountant's balance sheet, on one side, *the costs* (project displacement, waterlogging, salinity, sedimentation of the reservoir) as against, on the other, *the benefits* (irrigation, agricultural productivity, flood mitigation, hydropower). The strong, but not stated, assumption is that the CB can be credibly assessed by this simple tabulation of its measurable impacts. That is, there exists quantifiable 'facts' that can credibly establish a result. However, this CB ratio, rarely, if ever, is arrived at neatly (Porter 1995). Much of the confusion springs from the contested and political nature of how ecological values and prices are arrive at.[6]

Such a CB manoeuvre also presumes that one can cleanly or credibly transform ecological time into exact units of economic time. That is, all the costs and benefits are accounted for only within the time frame of the dam's estimated lifecycle. Thus, any long-term ecological consequences that could follow when the large dam's services (irrigation, hydroelectricity or navigation) terminate will not be treated as part of the immediate calculation. In other words, to assess the performance of a large dam, economic quantification must trump ecological process. The exercise of the 'quantifying spirit', that bends natural time to reveal economic value, in fact, becomes crucial to the translation of technical efficiency into quantifiable effects (see Lindqvist 1990).

---

[5]The literature on dam displacement in India is vast, but an excellent introduction to the subject is Dreze et al. (1997). For an argument that looks for good policy and best practices in order to turn resettlement and rehabilitation into potentially positive outcomes, see Mathur (2006).

[6]For an excellent critique of the neoclassical problem with valuing 'noneconomic' goods such as environmental relationships and the limitations of the cost-benefit-exercise, see Foster (1997).

For India, Satyajit Singh (1997) helpfully summed up some of the earliest questioning of the CB format. In an insightful review of several dam projects, he pointed out that the CB ratio was invariably a manipulated figure, in which the costs were made to move downwards while the benefits always tended to be overstated. The CB ratio, not surprisingly, has served as the dry gun powder for sparking many a resistance campaign against large dams in India. In the case of the controversial Sardar Sarovar Project (SSP), for example, Ranjit Dwivedi (2006) records how different CB outcomes were politically arrived at rather than being based on any objective criterion.[7]

The CB format, however, was flawed in other ways as well. A recent study by Radha D'Souza (2006c) points out that quantitative data on river hydrology were often conceptually suspect. In her study of the Krishna Water Disputes Tribunal (KWDT), D'Souza shows how attempts to 'scientifically' quantify hydraulic data were compromised not only by the fact that there was an absence of reliable time-series measurements on flows in the Krishna river but, interestingly enough, there existed sharply differing spatio-temporal scales in the reading of the river's hydrological cycle: between a geomorphological scale (river runoff and dependable flow), on the one hand, and a limited engineering scale (streamflow), on the other. Inevitably the selection of the data sets by the KWDT, D'Souza argues, was made as choices borne from political pragmatism rather than from any pretension to scientific judgement. These studies convincingly suggest that the CB ratio has been made to operate as a type of 'political arithmetic' in which the project is sought to be positioned as a neutral technological artefacts while all along being directed towards realising specific political outcomes. By subjecting the CB format to critical scrutiny, therefore, a new definition of the large dam is called for. For us, the large dam is the technical means to realise political outcomes. The dam enables the transfer of a region's hydraulic endowments to already empowered beneficiaries with the costs borne by dispossessed project-affected persons and marginal communities.

The hydraulic transfer is affected by the comprehensive transformation of the river's ecology. That is, the river is put to work by being altered into irrigation cusecs, kilowatts for hydroelectricity and as dead storage for flood control. The consequences following this dramatic overhaul in the river's character have been brilliantly discussed in a recent study by Shripad Dharmadhikary (2005). In *Unravelling Bhakra*, Dharmadhikary provides one of the most original discussions on both the CB format and the politics of the hydraulic transfer in India by reassessing the performance of the much celebrated Bhakra-Nangal Project (BNP). The BNP, made operational in 1963, comprises several dams, reservoirs, interbasin transfer linkages, powerhouses and a massive canal network intended to harness the waters of the Sutlej and Beas river (tributaries to the grand Indus River system).

[7]The agencies that carried out the CB for the SSP were the Narmada Planning Group (1983), Tata Economic Consultancy Services (1983), World Bank (1985), SSP Narmada Nigam ltd (1989) and again World Bank (1990) (see Dwivedi 2006).

For Dharmadhikary, the impacts of the BNP cannot be evaluated by a standard cost-benefit examination. He suggests that the project must be evaluated for creating win-lose scenarios rather than, as widely claimed, win-win outcomes. At the outset itself, the water availability for the BNP to irrigate 2.37 million hectares (mha), for example, was made possible by cutting off a near equivalent amount of supplies for 2.21 mha in the Sutlej Valley Project situated in Pakistan. Perennial canal irrigation, furthermore, was intended to initiate India's embrace of the Green Revolution agricultural strategy. The Green Revolution package, as it was popularly referred to, was aimed at producing a type of steroidal effect in agriculture. To that effect controlled and abundant irrigation became the means for facilitating a constellation of techniques and technologies that were intended to boost crop yields by profoundly reworking ownership and land tenure patterns (through consolidation), introducing new input packages (chemical fertilisers, high-yielding varieties, mechanisation) and encouraging crop monocultures.

The Green Revolution boost to yields in mostly cereal production, however, has been compromised by declines elsewhere. Dharmadhikary notes that waterlogging, salinisation and the deleterious effects on soils from intensive mono-cropping have plagued the canal-irrigated tracts, besides many of the famers have their profits squeezed by the scissor effect of mounting input costs and the tapering off in yields. Clearly, a simple CB format is inadequate for the task of capturing long-term ecological and economic trends. And as for the hydraulic transfer effected by the Bhakra-Nangal dams, Dharmadhikary argues that the project actually ended up amplifying earlier British colonial land and water management initiatives in the region, begun in the long nineteenth century, a period in which colonial policies led to the eradication of most cattle-rearing 'wandering tribes', the conversion of once forested tracts and grasslands into mono-cropped commercial wheat fields, the elimination of subsistence cultivator communities who depended on inundation irrigation and the oppressive exploitation of settled agriculturalists with colonial revenue demands.[8] Colonial attempts, hence, aimed to transform a variegated social and ecological flood plain into an administratively simplified settled agrarian tract (Dharmadhikary 2005).

The entire Indus River system, which straddles both contemporary India and Pakistan, has been substantially drained off its waters by a series of dams, barrages and canals systems. Historically, it has been calculated that before the great siphoning projects through perennial canal irrigation had begun in the latter half of the nineteenth century, up to 150 maf (million acre-feet) of fresh water probably flowed into the delta, along with the deposition of close to 400 million tons of nutrient-rich fertilising silt. The Indus delta was then a sprawling interstitial zone between land and sea and made up of mangroves, inlets, creeks and an inestimable number of ecological relationships between flora and fauna (Dharmadhikary 2005). But subsequent to the damming and diversion of the Indus and its tributaries for

[8]For a discussion on the precolonial ecological context of the semiarid Punjab plains, see Agnihotri (1996). Also see Bhattacharya (1996) and Singh (1991).

agriculture, power and nation-building, the amount of fresh water flowing into the Indus delta has been reduced to a lean 10 maf (less than 10 % of historical flows). The full effect of this massive siphoning off of fresh water from the delta has only now begun to be acknowledged. Besides debilitating livelihood possibilities for approximately 1.2 million people, who live in the delta and along the coasts, the fluvial impoverishment of the delta has resulted in tangible negative impacts on fish breeding, damage to marine food webs, destruction of unique salt water ecological habitats and an inestimable loss in biodiversity (WCU 2003).

The politics that effected hydraulic transfers in India have, however, undergone a further twist in recent years. Increasingly, large dams or MPRVD projects are now redirecting river flows for urban and industrial consumption. The brewing conflict over the apportionment of the waters of the Narmada River is one such clear instance. Originally intended to 'benefit' 29 million people across 8,215 villages and 135 towns in the drought-prone areas of Saurashtra, Kutch, North Gujarat and Panchmahal, the Gujarat Water Infrastructure Limited (GWIL) has piped the much-awaited flows instead to the city of Gandhinagar and oversupplied it to industries in Kutch (Bhattacharya 2007). In the state of Orissa (Eastern India), in November 2007, some 30,000 farmers gathered at the reservoir of the Hirakud dam (Sambalpur district). Surrounding the reservoir, they demanded that the government ensure that the waters be committed for irrigation rather than being directed towards industry. Despite the subsequent police action of arrests and beatings, the farmers remained firm in their resolve. In fact, 10 days after the protest, farmers reassembled to erect a 16-ft-long wall above an underground pipe laid by Vedanta Aluminium to move water from the reservoir to one of their smelting units. The wall was named *Chasi Rekha* (farmers' demarcator) and has since become the farmer's rallying symbol for asserting their rights over the reservoir (Mahapatra and Panda 2007).

While such kinds of hydraulic transfers are, undoubtedly, enabled by large dams, the ideological context for shifting and shuffling flows between different constituencies however draws upon the now troubled belief in supply-side hydrology. Simply put, supply-side hydrology, in the words of its enthusiasts, refers to a strategy wherein the need for 'an additional quantity of water' is met by increasing 'the available supply of water through new development projects' (Biswas and Embid 2003). Supply-side hydrology has meant that initiatives to ameliorate perceived shortages have been met either by the construction of dams and diversions or by encouraging groundwater mining through electric and diesel pumps (D'Souza 2003a). Such supply-side interventions, moreover, have usually advocated expert-driven, centralised, bureaucratically organised and a quantitative approach towards water manipulation. One of the grave consequences in such a top heavy and non-participatory model has been a pronounced lack of sensitivity to local complexity, especially in ignoring how water access is shaped by processes of political economy and competitive social interests. In a study Lyla Mehta (2005) has brilliantly argued that the idea of 'water scarcity' in Kutch (Gujarat) was as much a product of meteorological parsimony as it was a notion manufactured by several political and policy interventions. Mehta describes how, from the late 1960s on, various state discourses and programmes treated water shortages as a result of a natural limit. This notion of water as an absolute natural scarcity not only fuelled the rhetoric for the construction of the Narmada Sagar dam but simultaneously worked to marginalise local water knowledges and livelihood strategies of the nomadic, pastoral and subsistence peasants that had been creatively adapted to cope with seasonal scarcity. In other words, the notion of absolute scarcity, critical to the ideology of supply-side hydrology, was deployed to meet water demands through either big-engineering projects or intensive extraction technologies rather than concentrating on localised conservation efforts or on strengthening indigenous water knowledge traditions.

Water management in India is now dominantly controlled by centralised water bureaucracies, contractors, private engineering firms, finance houses (such as the World Bank and Asian Development Bank) and powerful political lobbies. These expert-led institutions and organisations with immense financial and political powers have systematically moved to either expropriate indigenous water techniques or cause the destruction of water management traditions.

4.3 Concluding Remarks

The large dam in India can no longer be treated as a political conundrum. As the above discussion argues, large dams, though announced as neutral technological artefacts, are deeply implicated in several processes that have been integral to effecting types of enclosure, hydraulic transfer, the expropriation and elimination of other water management skills and traditions and inevitably the externalisation of real costs through displacement onto the most marginal and impoverished communities.

Furthermore, the modern large dam must also be understood as being crucial to sustaining supply-side hydrology – the dominant modern paradigm for harnessing water in the contemporary period. In recent years, however, the pursuit of supply-side hydrology, the world over, has begun to flounder especially over the question of its environmental impacts. The triptych of strategies involving groundwater mining, perennial canal irrigation and large dams, in other words, has proved to be unsustainable as a water management framework. At heart is the growing realisation that civil engineering and bureaucratic framings of rivers, as merely moving masses of water crying out to be regulated and dammed, are flawed.

In sharp contrast to these highly simplified views, ecologists have convincingly demonstrated that fluvial regimes are complex geomorphologic, chemical and biological processes in motion. Rivers are made up of habitat mosaics that support a wide variety of aquatic and riparian species. And the beating heart that keeps alive the river's ecological health and viability is its *natural flow regime*, which organises and defines the river ecosystem itself. It is now understood that natural variable flows create and maintain particular dynamics between the channel, floodplain, wetland and the estuary. The magnitude and frequency of high and low flows, consequently, help regulate numerous ecological processes. While wetlands provide important nursery grounds for fish and export organic matter and organisms into the main channels, the scouring of floodplain soils by floods rejuvenates habitats for plant species within the basin. Even periods of low flow provide ecological benefits, through the recruitment of different plant species. A large body of evidence now shows that the natural flow regime of virtually all rivers is inherently variable, and that this variability is critical to ecosystem function and native biodiversity (Poff et al. 1997; also see Ward and Stanford 1995). In fact, a variety of communities in India continue to tap and sustainably harness the variable ebbs and pulses in flow regimes through a range of innovative livelihood traditions and subsistence strategies. These riverine or river-dependent communities, not unexpectedly, are invariably debilitated or impoverished when large dams or diversions affect river flows.[9] In the main, traditional fishing populations and subsistence cultivators, with their livelihoods tuned to distinct river ecologies, have suffered the most. Their plight, as yet, remains largely undocumented and their losses have rarely been accounted for.

Nevertheless, despite the paucity of credible documentation, enough cause for alarm has already been sounded over several aspects of supply-side hydrology. In one recent study, for example, strong evidence indicates that overexploitation of groundwater has led to the rapid depletion of water tables, saltwater encroachment, drying of aquifers and groundwater pollution in many parts of India. Acknowledged as well is the fact that water tables are declining at the rate of 1–2 m a year (Singh and Singh 2002). On the reverse, in intensive canal-irrigated tracts, waterlogging has emerged as a widespread phenomenon with water tables rising up to 1 m a year, often leading to soil salinisation. By the late 1980s, a period for which we have some reasonably accurate estimates, salinised soils in India were close to 7 mha, which added up to roughly 17 % of the total land that was then under canal irrigation (Postel 1999). At the same time, the impacts of large dams in India have singularly suffered from the lack of any credible official examination. If anything, the purported successes or failures of the large dam in India continue to remain a state secret, despite several independent studies and reports indicating that all is not well in their functioning.[10]

More than ever, the large dam and supply-side hydrology in India urgently awaits a political resolution. On one side are the contractors, private engineering firms and centralised water bureaucracies, who, in recent years, as a last gasp effort, have been advocating for the interlinking rivers project. This is essentially a business-as-usual model, in which 37 rivers in India are sought to be connected through 30 diversions or links and 36 major dams. On the other side of the fence are innumerable popular

[9]See the moving account by Maheshwari (n. d.) in a pamphlet on the fate of the untouchable castes living on the banks of the Godavari River and threatened by submergence following the government's decision to build the Polavaram dam.

[10]A fairly substantial number of criticisms against large dams in India have emerged from the 'non-expert'. One of the first authoritative non-official overviews of the negative impacts by large dams in India was Agarwal and Narain (1985); also see Agarwal et al. (1999). Two excellent recent collections on large dams in Northeast India are Saghal (2003) and Menon and Kohli (2005); also see Lahiri-Dutt and Wasson (2008).

movements, potential victims of displacement and a rising crescendo of voices that are now loudly arguing for the abandonment of the existing water paradigm. Much depends on which way the tide turns in this round.

References

Agarwal A, Narain S (1985) The state of India's environment, 1984–85: the second citizens' report. CSE, New Delhi (reprint 1996)
Agarwal A, Narain S (1997) Dying wisdom: rise, fall and potential of India's traditional water harvesting systems. CSE, New Delhi
Agarwal A, Narain S, Sen S (eds) (1999) The citizen's fifth report on environment in India. CSE, New Delhi
Agnihotri I (1996) Ecology, land use and colonisation: the canal colonies of Punjab. Indian Econ Soc Hist Rev 33(1):37–58
Bhattacharya N (1996) Pastoralists in a colonial world. In: Arnold D, Guha R (eds) Nature, culture, imperialism – essays on the environmental history of South Asia. Oxford University Press, New Delhi, pp 49–68
Bhattacharya DP (2007) Narmada water project leaves drought-prone areas thirsty. The Indian Express (2 April 2007). Online edition. http://cities.expressindia.com/fullstory.php?newsid=229515. Accessed 11 Feb 2011
Biggs D (2006) Reclamation nations: the U.S. Bureau of Reclamation's role in water management and nation building in the Mekong Valley, 1945–1975. Comp Technol Transf Soc 4(3):225–246
Biswas AK, Embid A (2003) Editorial. Int J Water Resour Dev 19(3):351–352
Dharmadhikary S (2005) Unravelling Bhakra: assessing the temple of resurgent India. Manthan Adhyayana Kendra, Badwani
Dreze J, Samson M, Singh S (1997) The dam and the nation: displacement and resettlement in the Narmada Valley. Oxford University Press, New Delhi
D'Souza R (2003a) Supply-side hydrology in India. The last gasp. Econ Pol Wkly 38(36):3785–3790
D'Souza R (2003b) Damming the Mahanadi River: the emergence of multi-purpose river valley development in India (1943–46). Indian Econ Soc Hist Rev 40(1):81–105
D'Souza R (2006a) Water in British India: the making of a 'colonial hydrology'. Hist Compass 4(4):621–628
D'Souza R (2006b) Drowned and dammed: colonial capitalism and flood control in Eastern India. Oxford University Press, New Delhi
D'Souza R (2006c) Interstate disputes over Krishna waters: law, science and imperialism. Orient Longman, Hyderabad
Dwivedi R (2006) Conflict and collective action: the Sardar Sarovar Project in India. Routledge, London
Foster J (ed) (1997) Valuing nature: economics, ethics and environment. Routledge, London
Hart HC (1956) New India's rivers. Orient Longman, Mumbai
Hoag HJ (2006) Transplanting the TVA? International contributions to postwar river development in Tanzania. Comp Technol Transf Soc 4(3):247–267
Klingensmith D (2007) "One valley and a thousand": dams, nationalism, and development. Oxford University Press, New Delhi
Lahiri-Dutt K, Wasson RJ (eds) (2008) Water first: issues and challenges for nations and communities in South Asia. Sage, New Delhi
Lindqvist S (1990) Labs in the woods: the quantification of technology during the late Enlightenment. In: Frangsmyer T, Heilbron J, Rider RE (eds) The quantifying spirit in the 18th century. University of California Press, Berkeley, pp 291–314

Mahapatra R, Panda RK (2007) Ground swell. Down Earth 12(2007):22–30
Maheshwari RU (n. d.) Godavari: yet another dam, yet another story. Peace, New Delhi
Mathur HM (ed) (2006) Managing resettlement in India: approaches, issues and experiences. Oxford University Press, New Delhi
Mehta L (2005) The politics and poetics of water: the naturalisation of scarcity in Western India. Orient Longman, Hyderabad
Menon M, Kohli K (eds) (2005) Large dams for hydropower in Northeast India: a dossier. SANDRP, Pune
Pearce F (2005) Dams and floods. In: Menon M, Kohli K (eds) Large dams for hydropower in Northeast India: a dossier. SANDRP, Pune, p 99
Poff NL, Allan JD, Bain MB, Karr JR, Prestegaard KL, Richter BD, Sparks RE, Stromberg JC (1997) The natural flow regime. BioScience 47(11):769–784
Porter TM (1995) Trust in numbers: the pursuit of objectivity in science and public life. Princeton University Press, Princeton
Postel S (1999) Pillars of sand. W. W. Norton, New York
Rao KL (1978) Cusecs candidate: memoirs of an engineer. Metropolitan, Kanpur
Roy A (1999) The cost of living. Flamingo, London
Sahgal B (ed) (2003) Large dams in Northeast India. Rivers, forests, people and power. The Ecologist Asia 11(1)
Scott JC (2006) High modernist social engineering: the case of the Tennessee Valley Authority. In: Rudolph LI, Jacobsen JK (eds) Experiencing the state. Oxford University Press, New Delhi, pp 3–52
Singh C (1991) Region and empire: Punjab in the seventeenth century. Oxford University Press, New Delhi
Singh S (1997) Taming the waters: the political economy of large dams in India. Oxford University Press, New Delhi
Singh P (2008) The colonial state, Zamindars and the politics of flood control in North Bihar (1850–1945). Indian Econ Soc Hist Rev 45(2):239–259
Singh DK, Singh AK (2002) Groundwater situation in India: problems and perspectives. Int J Water Resour Dev 18(4):563–580
Virilio P (2002) Ground zero. Verso Books, London
Ward JV, Stanford JA (1995) Ecological connectivity in alluvial river ecosystems and its disruption by flow regulation. Regul Rivers: Res Manage 11(1):105–119
WCD (World Commission on Dams) (2000) Dams and development. A new framework for decision-making. Earthscan, London
WCU (World Conservation Union) (2003) Indus delta, Pakistan: economic costs of reduction in freshwater flows (case studies in wetland valuation 5). WCU Pakistan Country Office, Karachi
Whitcombe E (1995) Environmental costs of irrigation in British India: waterlogging, salinity and malaria. In: Arnold D, Guha R (eds) Nature, culture, imperialism: essays on the environmental history of South Asia. Oxford University Press, New Delhi, pp 260–291
Zachariah B (2005) Developing India: an intellectual and social history. Oxford University Press, New Delhi

Chapter 5
Dams, Riparian Settlement and the Threat of Climate Change in a Dynamic Fluvial Environment

A Case Study of the Damodar River, India

Kumkum Bhattacharyya and Michael J. Wiley

Abstract The Damodar River, a subsystem of the Ganga, has always been flood prone. People as well as governments throughout the centuries have dealt with the caprices of this vital water resource, using structures such as embankments, weirs, barrages and dams. Post-independence the Damodar Valley Corporation (DVC) constructed four multipurpose dams to facilitate regional development and reduce flood hazards. Post-dam hydrographs show decreased monsoon discharges, reduced peak flow and a shifting of peak flow from July to August to September. Despite the DVC dams, the lower valley is still vulnerable to flooding, because the transport capacity of the river has also been reduced. As a result, the frequency of bank full events today is again similar to that observed in the pre-dam period. Because of the control structures once mobile channel *char* lands have been stabilised and permanently settled by Bangladeshi refugees. These new riparian communities are threatened by the fluvial environment. Changing patterns in riparian land use, fostered by alterations in flow regime, coupled with the long-term prospect of increased rain variability due to climate change, appear to be increasing the risk of rare but devastating floods in the Lower Damodar.

Keywords Char lands • Climate change • Communities • Dams • Damodar Valley Corporation • Fluvial environment • Hydrology • Geomorphology • Water resources

---

K. Bhattacharyya, Ph.D. (✉)
Department of Geography, University of Central Arkansas, 302 B Irby Hall, Conway, AR 72035, USA
e-mail: kumkumb@uca.edu

M.J. Wiley, Ph.D.
School of Natural Resources and Environment, University of Michigan, G 166 DANA, 440 Church St., Ann Arbor, MI 48109, USA
e-mail: mjwiley@umich.edu

M. Nüsser (ed.), *Large Dams in Asia*, Advances in Asian Human-Environmental Research, DOI 10.1007/978-94-007-2798-4_5,
© Springer Science+Business Media Dordrecht 2014

5.1 Introduction

Finding environmentally sustainable approaches to river management nowadays has become more urgent than ever before, particularly with regard to understanding the perennially controversial impact of dams both on the fluvial systems and on the socio-economic environment. The need to apply scientific knowledge for economic and social benefit has been strongly felt in developing countries (of the tropics) since the early 1940s, and worldwide dam management issues have increasingly captured scientific attention in recent decades (Williams and Wolman 1984; Petts 1984; WCD 2000; EP/EC 2000; Chin et al. 2002; Molle 2006; Wiley et al. 2008, 2010; Grantham et al. 2008; Bhattacharyya 2011).

River water was one of the first natural resources to be exploited, and almost all major rivers in decolonised Latin American, Asian and African countries are now controlled. In India as elsewhere, River Basin Organisations (RBOs) first arose as a mechanism for the integrated planning of large water development projects, aimed at meeting growing requirements for power, irrigation and industry. At the same time governments and RBOs promised technological relief from flood hazards, which had massive and tragic impacts on riparian communities on a regular basis. Offering the twin benefits of increased productivity and decreased environmental hazard, the integrated development of many river basins was subsequently pursued throughout the world (White 1977; Saha 1981; Saha and Barrow 1981). In newly constituted India, one of the first such RBOs, the Damodar Valley Corporation (DVC), was founded in 1948; it was modelled on the Tennessee Valley Authority (TVA) of the United States (Chandra 2003; Pangare et al. 2009). The unalloyed benefits of river control measures, however, are no longer taken for granted. Floods still wreak havoc (Kundzewicz and Kaczmarek 2000; Kundzewicz and Schellnhuber 2004) and troubling ecological consequences worldwide (Doyle et al. 2003). The debate surrounding this issue, unfortunately, has become extremely polarised. River control structures have thus become social, economic and political issues.

Within the last decade, in particular, river resource management policy has undergone a major paradigm shift (Pahl-Wostl et al. 2007). The older ideas of river basin planning gradually intermingled with watershed and ecosystem management approaches (Molle 2006; Wiley et al. 2008) and were adjusted accordingly. The European Union's Water Framework Directive (WFD) defines a new strategy for meeting human needs while protecting the environment that is relevant to other regions of the world (EP/EC 2000; Grantham et al. 2008). From this perspective, a review of the first Indian multipurpose river valley project and its human and environmental impact through dam construction is timely. This study reports on changes over approximately a 70-year period focusing primarily on hydrogeomorphological and social consequences of flow regime alterations and channel modifications in the Lower Damodar River valley. It also reviews from the social perspective both impressive gains in development since the DVC began its work and longer-term changes in flood hazard, particularly with an eye on future climate changes. Lessons learned from this case study may be helpful in guiding the planning and management of rivers of a similar nature in India and elsewhere.

It is offered as a contribution to the larger study of sustainable watershed planning and the search for ways to meet growing global water demands in an ecologically sustainable manner (EP/EC 2000; Wiley et al. 2008, 2010; Grantham et al. 2008).

5.2 The Damodar River System

The Damodar River is a subsystem of the Ganges River system of India. It rises in the Chotanagpur watershed approximately at 23° 37′ N and 84° 41′ E and the geographical boundary of the basin lies between 22° 15′ to 24° 30′ N and 84° 30′ to 88° 15′ E. The catchment area of the river is about 22,000 km^2 of which about 19,000 km^2 are in uplands and 3,000 km^2 in plains of a deltaic nature. The length of the river is 541 km approximately. The river slope is 1.86 m/km for the first 241 km, 0.57 m/km the next 167 km and 0.16 m/km in the last reach. The main tributaries are the Barakar, Tilaiya and Konar. Below the confluence of the Barakar and Damodar, there are a few minor tributaries such as the Nunia and Sali. Once the main distributaries were the Khari, Banka, Behula and Gangur, but now they are all disconnected, independent rivers. Near Palla the river takes a sharp southerly bend. Below Jamalpur, the river bifurcates into the Kanki-Mundeswari and the Amta channel and joins the Hooghly (also spelled Hugli) at Falta some 48.3 km south of Kolkata (also spelled Calcutta). This study is concerned with the reach of the Damodar River below the confluencing outlets of the Maithon and Panchet reservoirs and the old confluence point with the Hooghly River, as these two lower dams have an enormous impact on the downstream fluvial environment (Fig. 5.1).

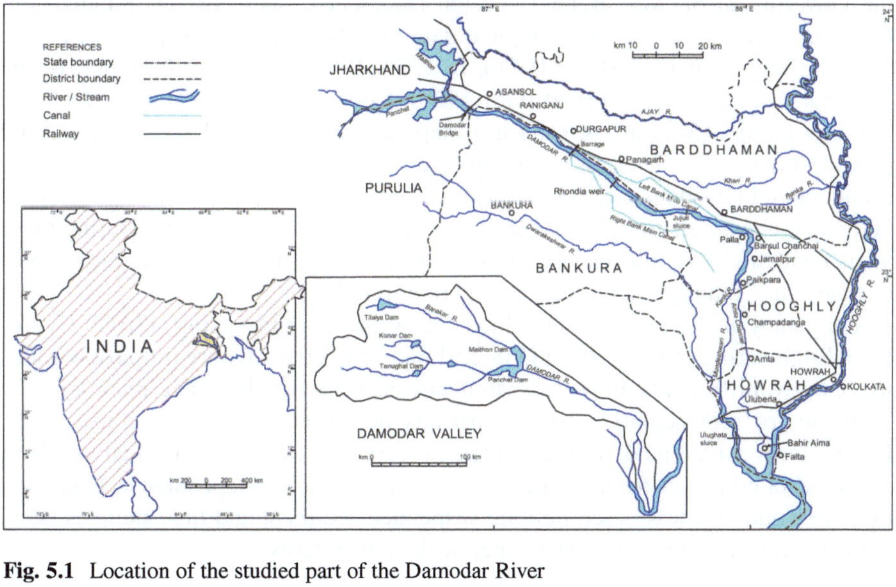

Fig. 5.1 Location of the studied part of the Damodar River

Above the reservoirs, the Damodar flows through the quartz-rich Archaean gneiss. Below, the Damodar flows through the sandstone-rich Gondwana sedimentaries near Asansol and Raniganj, which provide a rich source of fluvial sediments. The Damodar is a naturally flood-prone river. Gauge records at Rhondia (~95 km below Panchet; upstream catchment area) are extensive and relatively complete since 1933. Maximum flood discharge (recorded in 1913, 1935 and 1941) was 18,678 m^3/s at Rhondia. Average annual daily and minimum daily discharge for the period of record is 268 and 0 m^3/s, respectively. Floods in the river are the cumulative effect of rainfall on the catchment and the dynamics of catchment storage and channel transport. The nature of the river channel itself is especially important in this case, since aggrading beds and gradual shrinkage of cross-sectional areas and conveyance are increasing proneness to flood (Sen 1985a, b).

5.3 Methodology and Database

For a comparative analysis of the changes in the flow regime of the Lower Damodar River, data from 1933 to 2010 (according to availability) have been divided into two periods, the pre-dam period between 1933 and 1956 and the post-dam period between 1959 and 2010. It is worth mentioning that the lower dams Maithon and Panchet are fully functioning since 1957 and 1959, respectively.

Data were collected from different sources. To analyse the flow regime of the Damodar River, hydrological data from the following sources have been used: the Hydraulic Data Division, the Damodar Valley Corporation and the Central Water Commission, at Maithon, Jharkhand, and the Irrigation and Waterways Department, the Government of West Bengal and the Damodar Canals No. 11, Subdivision-Rhondia at Barddhaman. Hydrological data are also available in published form from the record of UNESCO (1971a, b, 1979, 1985) and from the report of the National Commission on Flood (GoI 1980). Geographical data sources included Survey of India (SoI) maps, cadastral maps and 2003 LISS-3 scenes of IRS-ID satellite image. The Dickens' map surveyed in 1854 (1:126,720) was also used. Frequent field visits were conducted to collect data on land use and riverbed landscape characteristics as these are significant issues in this study. During these visits residents were interviewed to obtain local perspectives on hydrogeomorphic changes, land management practices and river flooding.

Frequency analyses of flow and flood data were performed using USACE Hydrologic Engineering Center (HEC) DSS 2.01 and SSP 1.01 software (http://www.hec.usace.army.mil/). Daily flow data (summarised from hourly gauge records) were available for representative 10-year periods before and after dam construction. Longer-term annual peak flow data records were also available. We analysed both sets of data collected from the Rhondia gauging station independently. Dominant discharge analysis was based on total suspended sediment and flow samples taken at the Damodar Bridge site, located below the lower dams.

5.4 History of Flooding

In old records the Damodar has always been referred to as a river of sorrow. Flood propensity of the river is reflected in the construction of extensive embankments over 4,000 years ago as referred by hydraulic engineer Kapil Bhattacharyya (1959) and in sixteenth- to seventeenth-century maps (Fig. 5.2, Sen 1962; Bhattacharyya 1998, 2011).

The first recorded flood occurred in 1730 (Voorduin 1947). Floods with a peak flow of 8,496 m^3/s or more occurred 37 times between the years 1823 and 2007. The floods of 1823, 1840, 1913, 1935, 1941, 1958, 1959 and 1978 had peaks of more than 16,992 m^3/s. A peak flow of about 18,678 m^3/s was recorded in August 1913, August 1935 and again in October 1941 (DVC 1995; Bhattacharyya 2011). In an early report on flooding and embankments in this area by Sage et al. (1846), 13 severe inundations took place over a time span of 115 years, with seven inundations occurring in the first 85 years and six in the last 30 years. This increase in flooding may be attributed to the fact that, in the earlier period,

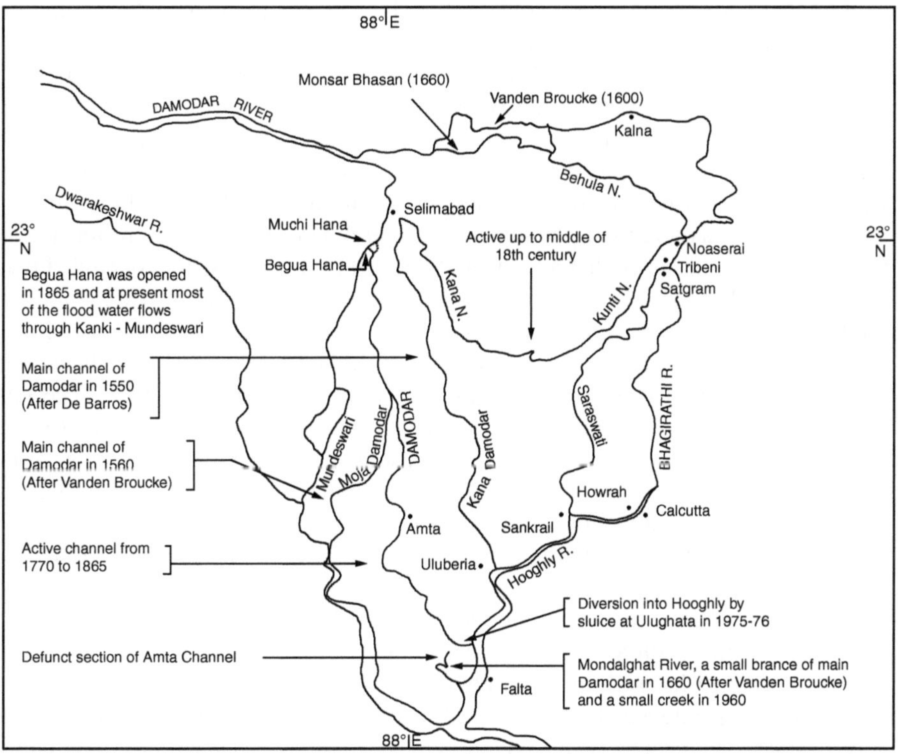

Fig. 5.2 Changing courses of the Damodar River (After Sen 1962; Basu 1989; Bhattacharyya 2002)

overflow irrigation from the Damodar was considered beneficial for agriculture, and people built and maintained canals to carry floodwater to their fields. From 1815 onwards, however, landlords and tenants of central Bengal started to neglect the irrigation canal systems, presumably beginning during the Maratha-Afghans war (1803–1818). The British thought these waterways were for navigation only and left them as they were. As these deteriorating waterways or canals took in lower amounts of water, more water remained in the Damodar main channel, and it grew to be a menace to the riparian tract (Willcocks 1930; Bhattacharyya 2011). Eventually the entire riverine regime of Bengal, Bihar and Orissa was transformed from a flood-enriched agrarian area into a landscape vulnerable to devastation by floods (D'Souza 2006).

5.5 Controlling the Lower Damodar

The flood-prone Damodar River has encouraged the construction of control structures from very ancient times; however, most traces have been destroyed completely due to the shifting of rivers (Fig. 5.2). The current embankments along the Damodar River were most probably constructed by local landlords to protect their land and property from floods (Gastrell 1863; Sengupta 1951) and certainly date from a period before British rule (O'Malley and Chakravarti 1909). The report of the embankment committee formed in 1846 states that these embankments were irregular and elevations were uncertain. The maintenance of embankments was inefficient and neglected (O'Malley and Chakravarti 1912), and breaching of embankments was a regular phenomenon. Several subcommittees were formed to investigate these issues (Bhattacharyya 1998, 2011). It was finally concluded that the removal of the right embankment for about 32.2 km would provide complete security to the left bank, which included the town of Barddhaman, the East India railway line and the populous districts of Hooghly and Barddhaman. In 1889 another 16.1 km of the right bank embankments was removed (Voorduin 1947). The left bank embankments were made continuous for a length of 176.87 km and were provided with many sluices. Attempts were also made to transfer excess water from the Damodar River to some of the decaying distributaries through the Eden canal in 1881. In 1933, the Damodar Canal system was opened. Water from the main river was admitted into the canal with the help of a weir at Rhondia near Panagarh of the district of Barddhaman.

Despite all of these measures, the river was flooding the lower sector throughout the British period (1908–1946). Concern about the devastation wreaked by the Damodar reached its peak in 1943, after one of the severest floods. Finally, after India won independence, the Damodar Valley Corporation (DVC) was founded on July 7, 1948. The project had originally envisaged the construction of eight dams and the planner estimated a design flood of 28,321 m^3/s with a 100-year frequency. To protect the lower valley, it was estimated that the design flood be moderated to 7,080 m^3/s, which was the total capacity of the Lower Damodar. However, due to

Table 5.1 DVC infrastructure – at a glance

DVC command area	24,235 km^2
Power management	
Total installed capacity	2,796.5 MW
Thermal power stations	4, Capacity: 2,570 MW
Hydel power stations	3, Capacity: 147.2 MW
Gas turbine station	1, Capacity: 82.5 MW
Substations and receiving stations	At 220 kV: 11 At 132 kV: 33
Transmission lines	At 220 kV: 1,342 km At 132 kV: 3,419 km
Water management	
Major dams and barrages	Tilaiya, Konar, Maithon and Panchet dams and Durgapur barrage
Irrigation command area (gross)	569,000 ha
Irrigation potential created	364,000 ha
Flood reserve capacity	1,292 million m^3
Canals	2,494 km
Soil conservation	
Forest, farms, upland and wasteland treatment	400,000 ha (approximately)
Check dams	16,000 (approximately)

Data source: DVC, Kolkata

financial and other constraints, the participating governments of West Bengal, Bihar (present Jharkhand) and the central government approved the construction of only four multipurpose dams (Voorduin 1947; Bhattacharyya 1998, 2011).

The DVC constructed the dams of Tilaiya and Maithon on the Barakar River, a tributary of Damodar, in 1952 and 1957, respectively; the Konar dam on the Konar tributary in 1955; and the Panchet dam on the Damodar River itself in 1959 (Tables 5.1 and 5.2). Maithon and Panchet act as control reservoirs and are located approximately 8 km above the confluence point of the Barakar and the Damodar Rivers. The construction of a barrage at Durgapur was started in 1952 and subsidiary structures were completed by 1958. The Tenughat Dam was constructed by the Government of Jharkhand in 1978 on the Damodar River; it was designed primarily for industrial and municipal water supply uses as well as for irrigation through the Tenu-Bokaro Canal.

5.5.1 Downstream Impact of Dams on the Hydrologic Regime

The retention of water behind a dam and its gradual release downstream result in the reduction of peak storm flows and an alteration of the entire flow and sediment

Table 5.2 DVC power plants – at a glance

Name	Location	Existing capacity	Commissioning
Hydel			
Tilaiya	River: Barakar Dist.: Hazaribagh State: Jharkhand	4 MW (2 × 2 MW)	U-I Feb. 1953 U-II Jul. 1953
Maithon	River: Barakar Dist.: Barddhaman State: West Bengal	63.2 MW (2 × 20 MW + 1 × 23.2 MW)	U-I Oct. 1957 U-II Mar. 1958 U-III Dec. 1958
Panchet	River: Damodar Dist.: Dhanbad State: Jharkhand	80 MW (2 × 40 MW)	U-I Dec. 1959 U-II Mar. 1991
Total hydel		**147.2 MW**	
Thermal			
Bokaro 'B'	Dist.: Bokaro State: Jharkhand	630 MW (3 × 210 MW)	U-I Mar. 1986 U-II Nov. 1990 U-III Aug. 1993
Chandrapur	Dist.: Bokaro State: Jharkhand	750 MW (3 × 130 MW + 3 × 120 MW)	U-I Oct. 1964 U-II May. 1965 U-III Jul. 1968 U-IV Mar. 1974 U-V Mar. 1975 U-VI Mar. 1979
Durgapur	Dist.: Barddhaman State: West Bengal	350 MW (1 × 140 MW + 1 × 210 MW)	U-III Dec. 1966 U-IV Sep. 1982
Mejia	Dist.: Bankura State: West Bengal	840 MW (4 × 210 MW)	U-1 Mar. 1996 U-II Mar. 1998 U-III Sept. 1999 U-IV Feb. 2005
Total thermal		**2,570 MW**	
Gas turbine			
Maithon	Dist.: Dhanbad State: West Bengal	82.5 MW (3 × 27.5 MW)	U-I Oct. 1989 U-II Oct. 1989 U-III Oct. 1989
Grand total		**2,799.7 MW**	

Data source: DVC, Kolkata

regimes (Williams and Wolman 1984; Petts 1984; Chien 1985; Kondolf 1997; Kondolf and Batalla 2006; Batalla et al. 2004; Graf 1999, 2006; Richter et al. 2010). The Damodar River was no exception. Based on gauging at Rhondia, before the DVC dams were constructed, about 35 % of the annual rainfall on the upstream basin was carried as flow through the lower river (Fig. 5.3, Table 5.2).

Since construction in 1957–1958, river flow at Rhondia has been substantially reduced to on average 23 % of catchment precipitation by both increased evaporation from the reservoirs and increased consumptive use of the stored water during dry periods. The representative daily average streamflow for the Damodar at

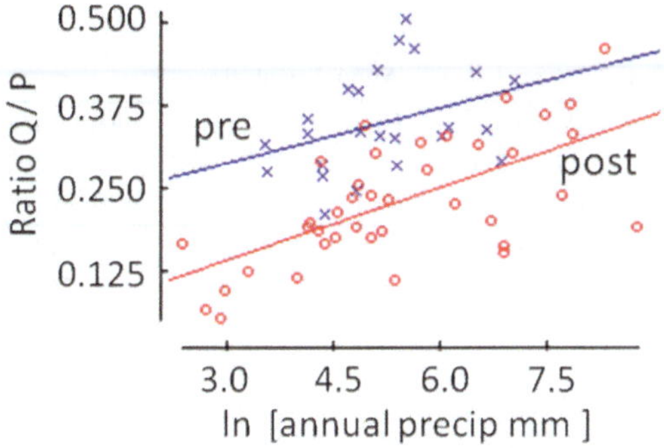

Fig. 5.3 Proportion of annual rainfall delivered as river flow at Rhondia before and after DVC dam construction (Data source: DVC, Maithon) (Note: More than a century of rainfall data from a large number of gauges distributed within the Upper Damodar Valley, which includes the Maithon and Panchet sub-catchments, are available from DVC, Maithon. Rainfall data for the monsoon seasons between 1891 and 2007 for both sub-catchments has been assembled and analysed)

Table 5.3 Summary of pre- and post-dam construction hydrology as measured at Rhondia

		Annual precipitation (mm)	Average daily flow (m^3/s)	Q/P ratio	Annual peak Q (m^3/s)	Return interval (years)	Dominant Q (m^3/s)	Bank full return (years)
Pre-dam (before 1957)	Mean	1,118	327	0.35	8,655	2.32	2,178	1.03
	SD	187	97	0.08	3,977	–	–	–
	Min	843	172	0.21	1,714	1.01	–	–
	Max	1,660	492	0.51	18,406	>500	–	–
	Years	67	24	24	30	–	–	–
Post-dam (after 1958)	Mean	1,155	234	0.23	3,646	1.36	720	1.06
	SD	259	128	0.09	2,321	–	–	–
	Min	644	32	0.05	344	1.01	–	–
	Max	1,714	623	0.46	10,919	231	–	–
	Years	47	42	39	53	–	–	–
	Contrast	NS	<.001	<.001	–	–	–	–

Upstream catchment: 8,837 km^2 (approximately). Ratio of cumulative annual river flow to cumulative catchment precipitation (Q/P) reflects proportion of precipitation delivered annually as river flow at Rhondia. Estimated dominant discharge is the expected equilibrium channel capacity at bank full and was calculated at the discharge from daily flow records and pre- and post-dam sediment loading functions (see Fig. 5.4)

Rhondia was 327 m^3/s in the pre-dam and 234 m^3/s in the post-dam period. Overall, the flow regime after the construction of the dam had higher baseflows but lower storm flows (Table 5.3), reflecting the additional storage capacity of the reservoirs and the reduced upstream catchment yields.

Average annual flood flows likewise declined (58 %) after the DVC projects were completed although there were again large interannual differences driven by variation in monsoonal precipitation. Frequency analysis (Table 5.4) suggests

Table 5.4 Frequency analyses for pre- and post-dam flow at Rhondia gauge

Daily average flow			Annual peak flow			
EF (%)	Pre-dam 1940–1950 (m^3/s)	Post-dam 1993–2008 (m^3/s)	EF (%)	Pre-dam 1933–1957 (m^3/s)	Post-dam 1959–2010 (m^3/s)	Return time (years)
1	3,575	2,460	0.2	17,682	11,542	500
5	1,977	1,139	0.5	16,950	10,742	200
10	1,263	681	1	16,271	10,029	100
20	577	322	2	15,456	9,209	50
30	200	168	5	14,104	7,926	20
40	85	99	10	12,797	6,772	10
50	38	66	20	11,129	5,415	5
60	19	37	50	79,126	3,164	2
70	9	16	80	50,529	1,577	1.25
80	3	6	90	3,816	1,024	1.11
85	0.7	2	95	2,9517	691	1.05

Data source: Damodar Canals No. 11, Subdivision-Rhondia, Barddhaman; GoI 1980; UNESCO 1971a, b, 1979, 1985

EF exceedance frequency (% of time the flow indicated is equaled or exceeded in the period of record)

Note: The daily streamflow (discharge) data from 1940 to 1949 at Rhondia, situated approximately 95 km downstream of Panchet Dam and 21 km from the Durgapur barrage, has been collected from the Damodar Canals No. 11, Subdivision-Rhondia, and the same data for 1993 to 2007 has been computed from hourly streamflow data available from the same location. The monthly average and minimum streamflow of the Damodar River at Rhondia from 1934 to 1979 have been collected from UNESCO (1971a, b, 1979, 1985). Data for the years 1961 to 1964, 1968, and 1975 are not available. The monthly average streamflow of the river at Rhondia for the period 1982–2007 has been calculated from the daily streamflow available from the Damodar Canals No. 11, Subdivision-Rhondia at Barddhaman. The annual peak streamflow data for the River Damodar at Rhondia are available from 1823 to 1933 as a discontinuous series from GoI (1980). The same data from 1934 to 1960 are published by UNESCO (1971a, b) and from 1960 to 2010 are available from the Damodar Canals No. 11, Subdivision-Rhondia, Barddhaman

that at Rhondia flood magnitudes of 8,000 m^3 and larger, once common (~2-year recurrence interval), have become relatively rare (<15 recurrence interval). And flood flows >12,000 m^3 have been entirely eliminated to date (pre-dam recurrence interval <10 years). In terms of reducing annual flood flow magnitudes, the DVC projects have been extremely successful. The capacity of the DVC system to moderate flood flows is impressive and is summarised in Table 5.5.

The discharge of the Damodar at Rhondia for four climatologic seasons with respect to the total annual flow from 1934 to 2007 is given in Table 5.6. The most important characteristic to note is that there has been at least a 30 % decrease in average annual discharge from the pre-dam to the post-dam period but also a nearly 48 % increase in flow variability. In contrast to the reductions seen during the monsoon, winter and summer flows have been substantially increased, and their variability has declined. After the dam closure monsoon water has been stored and

Table 5.5 Performance of Damodar Valley Reservoirs (Maithon and Panchet) in combined flood moderation during major flood periods[a]

Periods	Combined peak inflow (1,000 m^3/s)	Moderated outflow (1,000 m^3/s)	Flood moderation (1,000 m^3/s)	Flood moderation achieved
Sep. 1958	15.7	5.0	10.7	68
Oct. 1959	17.6	8.2	9.5	53
Oct. 1960	9.9	2.6	7.2	73
Oct. 1961	14.6	4.6	10.1	68
Jul. 1963	12.8	3.4	9.4	73
Oct. 1963	13.2	2.6	10.6	80
Sep. 1971	12.0	5.1	6.9	58
Sep. 1973	16.7	5.0	11.7	70
Sep. 1978	21.9	4.6	17.3	79
Sep. 1995	17.5	7.1	10.4	59
Sep. 1999	10.3	3.4	6.9	67
Sep. 2000	10.8	5.7	5.3	47
Sep. 2006	14.4	6.9	7.5	52
Sep. 2007	11.1	7.5	3.6	32

Data source: DVC 1995, MRO office, DVC Maithon

[a]Examination of maximum inflow and maximum outflow data for the two lower dams at Maithon and Panchet shows that flood moderation has been achieved during major flood years. Detailed examination of flow data, as available at Rhondia, revealed that maximum flow of 18,408 m^3/s had occurred twice in August 1913 and 1935 before the implementation of the DVC, but data shows that major floods nearing or exceeding this maximum observed flood of 18,406 m^3/s occurred during 1959, 1978 and 1995 and were successfully checked

Table 5.6 Seasonal streamflow characteristics of Damodar at Rhondia in pre-dam (1934–1956) and post-dam (1959–2007) periods (% of streamflow with respect to annual average streamflow total)

Period	Parameter	Summer	Monsoon	Autumn	Winter	Mean annual total (m^3/s)
Pre-dam	Years	21	21	21	21	–
	X	1.40	83.70	12.23	2.60	4,061.05
	SD	1.32	7.04	6.49	2.27	1,186.90
	CV	94.29	8.41	53.07	87.31	29.23
Post-dam	Years	48	48	48	48	–
	X	5.07	75.58	14.60	5.22	2,836.02
	SD	6.67	13.81	12.11	4.58	1,587.34
	CV	131.56	18.27	82.95	87.74	55.97

Data source: Damodar Canals No.11, Subdivision-Rhondia, Barddhaman
X average % of flow, *SD* standard deviation, *CV* coefficient of variation
Summer Mar–May, *Monsoon* Jun–Sep, *Autumn* Oct–Nov, *Winter* Dec–Feb

released during the non-monsoonal periods (i.e. summer, autumn and winter) to the Durgapur barrage, from which water is diverted to irrigation canals. Increasing demand for irrigation water has further enhanced the non-monsoon flow to some extent. Thus, there is considerable augmentation of flow in the lean season to

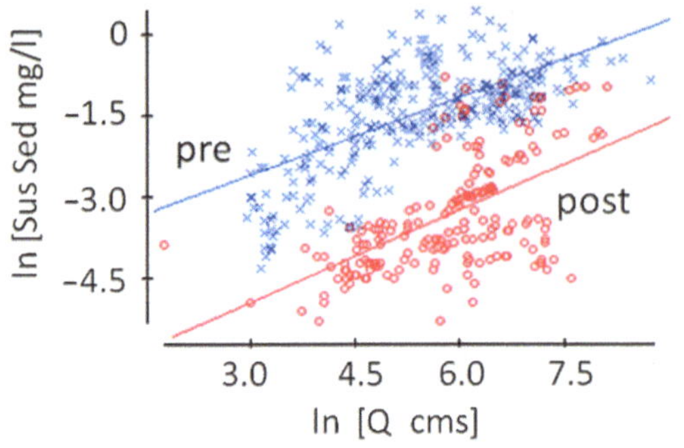

Fig. 5.4 Suspended sediment load pre- and post-dam construction at Damodar Bridge (Data source: DVC, Maithon) (Note: Information about suspended sediment concentration for the period 1952–1955 and 1981, 1984, 1985, 1986, 1987, 1988, 1991 and 1996–2001 have been computed from daily average suspended sediment concentration data provided by the Hydraulic Data Division DVC, Maithon)

meet the non-monsoon demand for irrigation. An irrigation system dependent on rainfall and river discharge has been replaced by more dependable irrigation through regulated release from reservoirs.

5.5.2 Changing Channel Morphology and Sandbar Formation

As the hydrologic conditions in the Damodar catchment have been modified through the development of control structures, the lower river channel system itself has been radically changed. Morphological changes along the Damodar River post DVC dam construction include both severe aggradation and narrowing of the channel through lateral accretion. Since the construction of the dam, the Damodar channel has undergone a general narrowing due to decreases in the flow and reductions in dominant discharge (Table 5.2, Fig. 5.4). The channel bed has been aggraded at an alarming rate in some places, whereas the volume of sediment and rate of sedimentation have grown in other areas. Although sediment is trapped in the reservoirs, the river still receives a million tons of sediment from the uncontrolled sectors below the dams in the Raniganj coalfield region. The declining capacity of the river to transport this sediment has made the channel bed a sediment sink with a downstream series of growing sandbars (locally known as *char* lands or *mana*). This effect is further enhanced by the coarsening of bed material. In some sections the Damodar channel has been reduced due to excessive sediment deposition following the flood of 1978. Channel reduction appears to have been achieved by the accumulation of sediment as shoals that are now vegetated and stabilised with agricultural fields and human settlements.

Sand loads carried by the Damodar have always given rise to sandbars within the river channel. Surveys by Captain C. H. Dickens' (1853) and SoI maps surveyed between 1929 and 1930 show that back then several semi-transient mid-channel and semi-mobile marginal bars covered with grass jungles existed. From 1947 onwards, after the partition of India and the former East Pakistan (present Bangladesh), refugees from Bangladesh began to occupy the sandbars. Encroachment on the *char* lands has continued to the present day due to the more hydrologically secure riparian environment created by multipurpose DVC projects. These anthropogenic landforms have acquired their present shapes and can be identified after construction of the Durgapur barrage (1958) and the Maithon (1957) and Panchet (1959) reservoirs. Some sandbars have merged with main land and most of the sandbars are populated. From the topographical sheets surveyed from 1985 to 1986, 2003 LISS-3 scenes of IRS-ID satellite image and from field visits during the period of 2007 to 2008, it is evident that new sandbars have emerged, although some have also been destroyed by channel migration.

Today, between the Maithon and Panchet reservoirs and the Falta outfall 250 km downstream, there are 23 larger occupied sandbars or *char* lands and a series of point bars. These *char* lands are now used as a resource base mostly by refugees who have matched land use at fine scales to flood experiences, applying their own concept of flood zoning to the riverbed and effectively assessing short-term and long-term risks.

5.5.3 *Social Consequences of Dams: Examples from Human-Induced Fluvial Environment*

Human alterations of the river's flow regime have resulted in the emergence of a series of stable sandbars that now characterise the Damodar River below the Maithon and Panchet dams. The local people did not identify the resource potentialities of these sandbars, since historically they were inundated and geomorphically ephemeral. These char lands remained as a stock of potential settlement sites. Refugees analysed the 'new' land more positively and responded rationally by considering it as a boon to their independent existence. The riverbed landscape at present is a result of functional relations that have developed between the riverbed and its occupiers (Bhattacharyya 1998, 2011). West Bengal received a large number of refugees after Indian independence with a fresh influx of Bangladeshi refugees after the Bangladesh War of 1970. In both phases a significant number of Bangladeshi refugees selected the riverbeds of West Bengal, including those of the Lower Damodar, as their second home. To illustrate the types of interactions that have occurred between river processes and riparian communities, we focus here briefly on two specific examples (Fig. 5.5).

Examples: Ramakrishna Palli, Pallishri and Sitarampur Mana

The Ramakrishna Palli, Pallishri squatters' colony and Sitarampur Mana are sited on a marginal but complex sandbar system below the Durgapur barrage. Pallishri

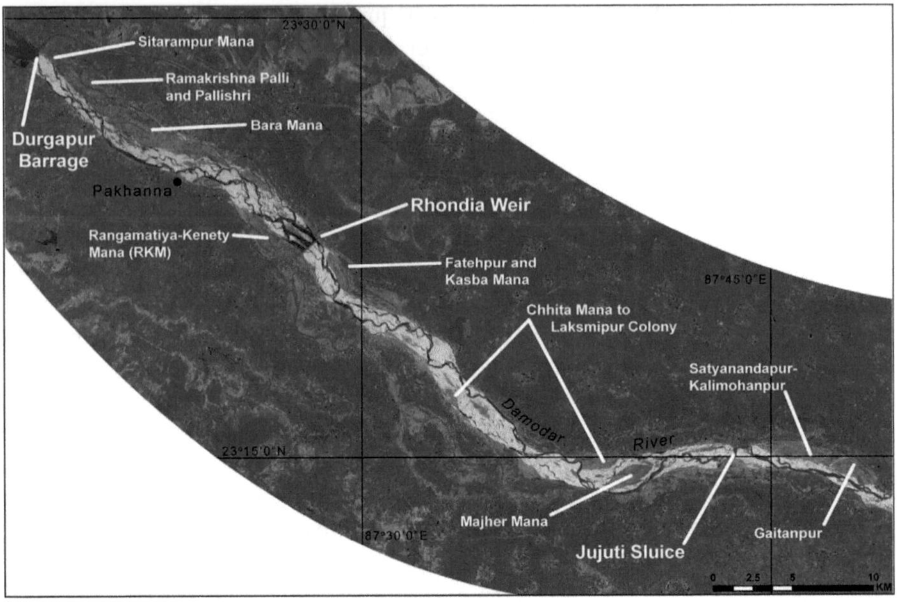

Fig. 5.5 Distribution of the sandbars below the Durgapur barrage showing the specific sandbars studied (Details of satellite image: Landsat ETM+; WRS2 Path: 139, Row: 04; Date of Pass: 19 March 2008)

used to be a squatters' colony but the residents have now been granted land deeds. Rivers are natural boundaries often taken as demarcating lines between countries, states and districts. Political boundaries usually remain static within a short time horizon and may be used to identify the shifting nature of a river.

Changes in generalised land use characteristics and landscape can be observed (Fig. 5.6). The bar material varies from fresh sand to clay; the clay deposits occur near former and current channel positions (e.g. D_2, D_3, D_4). Clay-bearing soils are significant, since they facilitate rice paddy culture. Multiple cropping is a common practice in this alluvial bar, and the dried-up portion of D_3 is used for double cropping since clay deposits of the dried-up channel are good for cultivating rice. This indicates that the people are aware of the deposition of clay materials in decaying channels and the higher field capacity of clay soil. Additional crops are grown on the bed of D_4 and in adjacent areas. These are unauthorised agricultural fields. Almost all types of vegetables, oilseeds and cereals are grown on these sandbars. Comparatively infertile lands are used for the cultivation of a coarse variety of jute (*mesta*). The oilseeds grown on these sandbars are mustard and sesame. Cereals like rice, wheat and many types of vegetables are also grown here.

Figure 5.7 shows hydromorphological characteristics, land use, and landscape features that include the channels of D_1, D_2, D_3, D_4 and D_5. The emergence of new sandbars and areas of double and multiple cropping are depicted. The main Pallishri settlement is located further from the channel D_1 at a higher elevation. This part

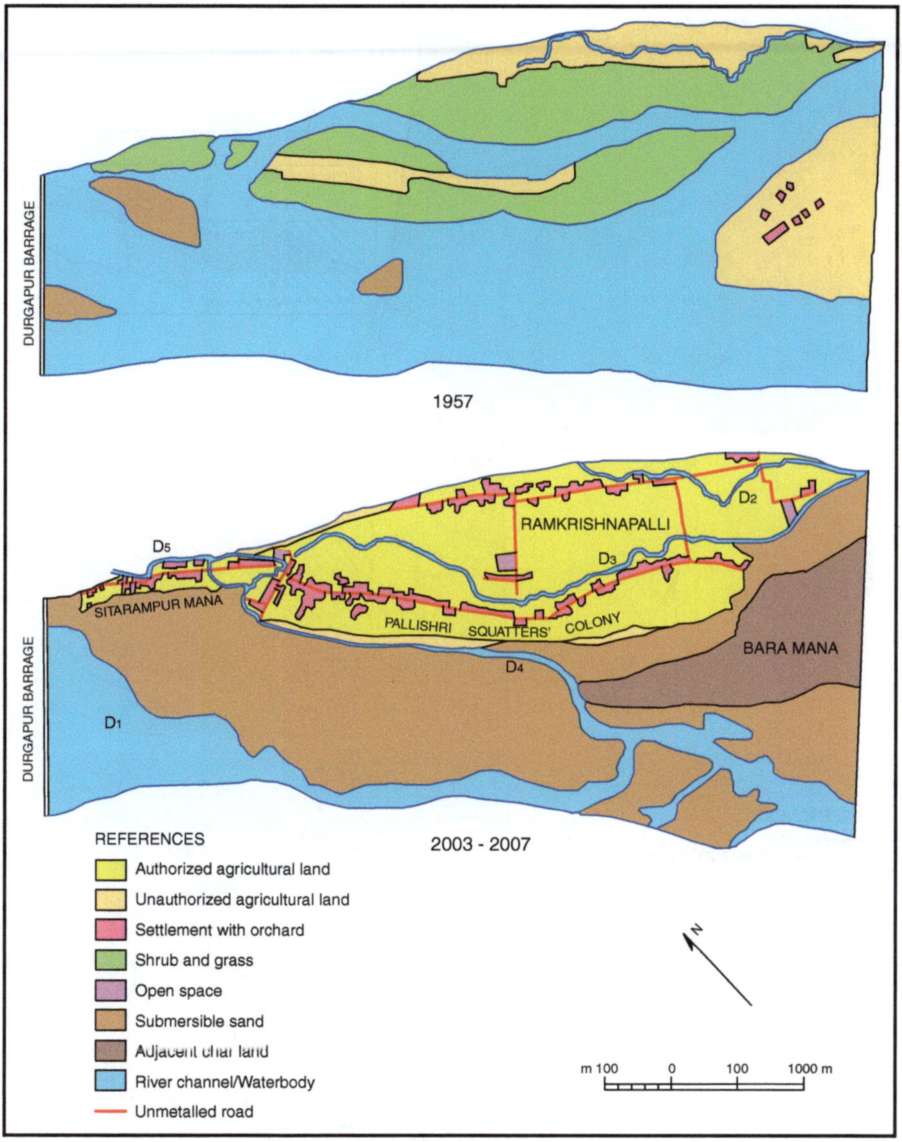

Fig. 5.6 Sitarampur Mana, Ramkrishna Palli, and Pallishri squatters' colony: changes in generalised land use characteristics and landscape (Note: Map prepared from cadastral or Mouza maps, Survey of India (SoI) maps of 73M/7N.W. (1:25,000), 73M/7 (1:50,000), 2003 LISS-3 scenes of IRS-ID and layout plans prepared between 1996 and 1998 and modified in 2007. Details of satellite image: IRS-ID, Sensor: LISS-3, Resolution: 23.5 m, Date of Pass: 23 January 2003, Path: 107, Row: 55, Bands: Green, Red, NIR & SWIR)

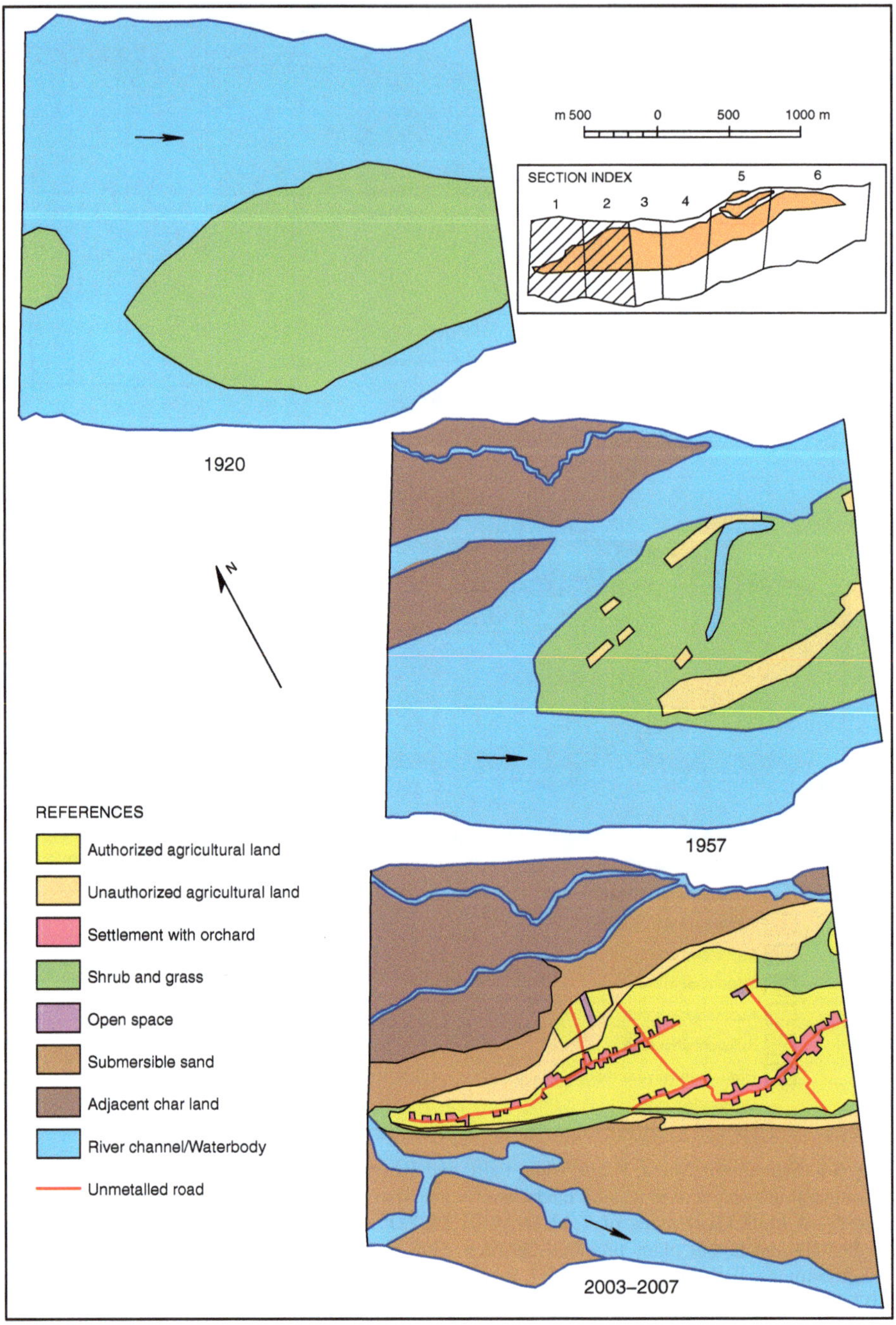
m 500
0
500
1000 m
SECTION INDEX
1
2
3
4
5
6
1920
N
1957
2003–2007
REFERENCES
Authorized agricultural land
Unauthorized agricultural land
Settlement with orchard
Shrub and grass
Open space
Submersible sand
Adjacent char land
River channel/Waterbody
Unmetalled road

of Pallishri, incidentally, has been the most stable part of the sandbar since 1957. Ramkrishna Palli is located close to the district boundary. The Sitarampur Mana is an extension of the Pallishri colony. The village roads are running almost parallel. In fact, village roads follow the boundary line between agricultural plots, particularly towards the south-western part of Pallishri.

The current riverbed landscape shows diversity at micro-level. Sociocultural, economic and political backgrounds of the riverbed communities are reflected in land use practices. The identification of resources and their utilisation and conservation depend on resource perception. The resource potentialities of the *char* lands were not perceived initially by the local people. This remained as a stock of fundamental geomorphic entities. The perception of the migrant communities has been controlled by their low position in the social space. Land use in the *char* lands follows the dictate of micro-relief on the one hand and perceived environment of the riverbed occupiers on the other (Bhattacharyya 1998, 1999, 2011).

Example: Bara Mana

Bara Mana, the largest alluvial sandbar in the culturally defined Lower Damodar, is sited below the Durgapur barrage. To review the changes in generalised land use characteristics and landscapes, Bara Mana has been divided into six sections, two of which are shown in Fig. 5.7. Over the last decades, the grass-covered land has been turned into agricultural fields. On the cadastral maps of 1957, some unauthorised agricultural fields can be found; permanent settlements developed after 1958 and 1959, upon completion of Durgapur barrage and Maithon and Panchet reservoirs. Section 1 comprising Bamandihi Mouza is shown in Fig. 5.8, which served as reference for the following observations:

1. Grasses are allowed to grow on the erosion-prone bank.
2. Jute is cultivated in the adjacent zone.
3. Double cropping is a common feature on the extreme west of Bamandihi.
4. Additional crops are grown in the unauthorised plots towards the north.
5. Multiple cropping is practised on higher land above the inundation level.
6. Linear settlements are to be found on higher ground in the flood-free area.
7. Linear settlements of low density are observed in inundation-prone areas towards the west. This part of the bar is also devoid of irrigation facilities.

Fig. 5.7 Bara Mana (*sections 1 and 2*), changes in generalised land use characteristics and landscape (Bamandihi and Purakonda Mouza) (Note: Maps prepared from cadastral maps, 2003 LISS-3 scenes of IRS-ID satellite of Indian National Remote Sensing Agency (*NRSA*), Hyderabad. Cadastral Survey (*CS*) maps (Bamandihi, Purakonda), surveyed between 1917 and 1924, and Revision Survey (*RS*) maps surveyed between 1954 and 1957 and several layout plans originally prepared between 1996 and 1997 and modified between 2007 and 2008 have been used. Mouza maps have been collected from District Collectorates. Survey of India (*SoI*) maps of 73M/7N.E., 73M/7N.W., 73M/7S.E. (1:25,000), 73M/7 (1:63,360, 1:50,000), IRS Geo-coded imagery 73M/7 and satellite image of 1994 IRS IB LISS-2 FCC classified image 1:100,000 have been consulted)

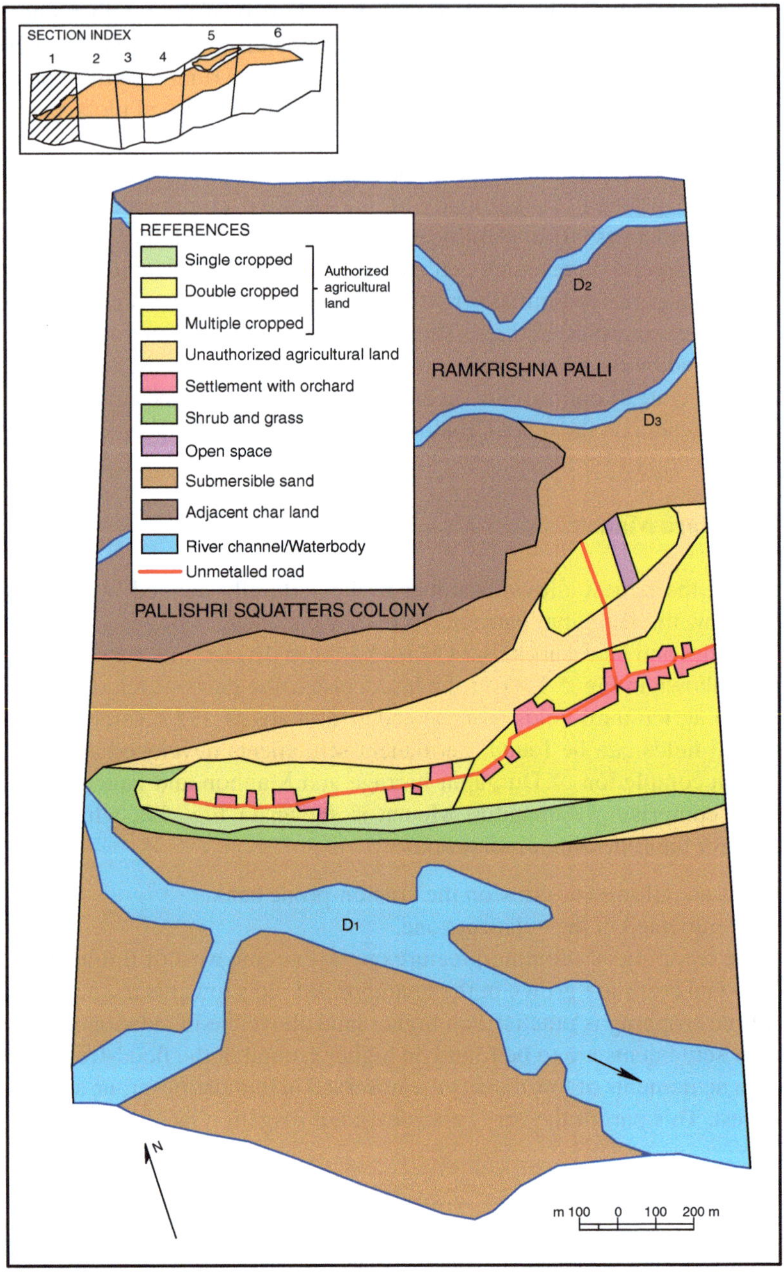

Fig. 5.8 Bara Mana (section 1) land use characteristics Bamandihi Mouza (Note: Map prepared from layout plans of Bamandihi Mouza prepared between 1996 and 1998 and modified in 2007 and through active field survey)

Early on, migrants began to perceive the agricultural potential of the sandbars as they came originally from farm sectors adopting themselves to similar tracts with wet rice culture. Bolai Paik, a well-known farmer from Bara Mana, reported that farmers first used this tract as grazing grounds. Falling leaves and animal droppings enhanced the fertility of the sandbars. Land use on the sandbars follows the dictate of micro-relief. The crop calendar is finely adjusted to match the flood-prone micro-environment. In higher parts of the stable sandbars above inundation level, settlement sites are found with individual huts on a higher plinth. The settlements are aligned strikingly linear along the crest of the convex bars. In these higher parts of the bars, perennial tree crops have been introduced, as well as floriculture and silkworm rearing mulberry cultivation. The zone between the higher tract and the inundation level is used for cultivating cash crops, cereals and vegetables. Inundation-prone areas are used to grow additional crops. Where habitation is not possible due to the low height of the bars, sand quarrying has become a dominant activity. Moreover, settlers often take the risk of growing capital-intensive vegetables even in the most vulnerable tract, as ready markets are present in Barddhaman and Durgapur. On these sandbars the groundwater table is high and a crop is assured with irrigation through shallow pumps. Present land uses such as the site and structural pattern of settlements, site and selection of crops and mode of irrigation water are very carefully attuned to the best utilisation of the local resource base (Basu and Bhattacharyya 1991; Bhattacharyya 1998, 2011).

5.6 Discussions and Conclusions

The natural resources and landscapes of the planet nowadays are controlled mostly by anthropogenic forces rather than natural forces (Turner et al. 1990; Messerli et al. 2000), a dynamic that has led to the suggestion that the current geological epoch be renamed 'Anthropocene' in place of 'Holocene' (Crutzen and Stoermer 2000; Ehlers 2008 cited in Baghel and Nüsser 2010; Zalasiewicz et al. 2008). The Damodar River has been controlled by government and hydraulic communities throughout the centuries, by employing both macro- and microscale planning and variously scaled river training programmes: embankments, canals, sluices, weirs, dykes, barrages, dams and reservoirs are now viable components of a historically conditioned geomorphic landscape (Bhattacharyya 1998, 2011). The Damodar's geo-fluvial landscape, therefore, like other major rivers of the world today, is largely anthropogenic in character.

We believe that the Damodar River control project has been clearly beneficial in terms of regional, economic and social development (Tables 5.1 and 5.2). The thermal power plants supply electricity to the region and have encouraged substantial industrial growth in nearby cities. Irrigation has also been drastically increased (Bhattacharyya 2002). Human settlers, taking advantage of reduced flooding and newly emerging *char* lands, have further augmented the resource potential of the river by building local control structures and carefully planning their own land use strategies (Bhattacharyya 1999, 2011). For the riverbed settlers of the

Damodar River, as noted by Wolman (2007, pers. com.), '…land use, cropping and settlement patterns, at a micro level, represent a sophisticated understanding of the flood regime and opportunities to maximise the productive capacity of an unpromising environment'.

However, while the promise of economic development appears to have been realised by river control, the promise of reduced flood hazard has proven more elusive. Geomorphological processes ensured that channel shape configurations would over time be adjusted to the new hydrologic regime initiated by the DVC developments. Today, although the magnitudes of annual floods have been reduced significantly, the frequency of over-bank flooding is approaching that of the pre-dam period. Channel aggradation, driven by flow reductions, has produced a new equilibrium between flow and sediment loading which retains the much observed relationship between channel capacity, flooding and annual floods with a return frequency of 1.2–1.5 years. However, years of flood suppression following the construction of the reservoirs and the predictably reduced flow rates have encouraged wholesale settlement of *mana* and in recent years even encroachment in the main channel itself.[1] As a result, the question must be whether the risks of catastrophic flooding losses have in fact increased in the long run as a result of what was originally conceived of as a flood control project. Riparian communities today find themselves in a clearly more vulnerable position when large events occur (e.g. 1978) than ever before.

Given that to assess the flood hazard both frequency and vulnerability have to be taken into consideration, it is interesting to ask what role the past 50 years of flood suppression may have played in promoting future vulnerability. Linking considerations of the hydrologic, geomorphic, and social consequences of dam construction can lead to a very different assessment of the potential hazards than using the more traditional focus on flood hydrology alone. This clearly is an argument for the kind of integrated water resources management perspective now being promoted across much of the WFD (see EP/EC 2000; Grantham et al. 2008; Pangare et al. 2009). Reviewing the history of floods and flood control in the Lower Damodar, it is clear that doing away entirely with floods is neither possible nor desirable. The goal should be not to eliminate floods but to incorporate them within our water management strategy, as with 'flushing floods' (DVC 1957) or Environmental Flow Requirements (EFRs), which deal with the amount, timing and conditions under which water should be discharged from the dams to retain the natural integrity of the downstream river ecosystem (Bergkamp et al. 2000). Controlled releases could be maintained without excessive flooding of the populated *char* lands with the goal of sustaining sufficient channel capacity to carry many common floods of frequent recurrence.[2]

---

[1]Biswas, P. Personal communication through phone conversation, dated 7 May 2011, a farmer of Bara Mana, located at the Damodar River.

[2]Wolman, M. G. Personal communication through letter dated 10 April 2006 and an e-mail/letter dated 11 January 2007, Late Professor, Department of Geography and Environmental Engineering, The Johns Hopkins University, Baltimore, MD, USA, pp 21218–2686.

The most crucial problem in the Lower Damodar, however, is that of anthropogenic encroachment and attempts at stabilisation of the active riverbed by the riverbed occupiers. Encroachment can both contribute to higher flood peaks by reducing hydrologic storage (Kundzewicz and Kaczmarek 2000) and increase the damages caused by flooding by increasing the population's vulnerability. The fact that most of the riverbed occupiers are Bangladeshi refugees makes this a critical regional and political issue. Increasing demand for living space and food creates a pressure to utilise hazard zones, including flood-prone areas (Bird 1980). So it is not unlikely that riverbeds in India and other countries, particularly in tropical countries, may be occupied by permanent settlements and transient sandbars may become immobile causing channel deterioration. These are problems that were not anticipated when the DVC was planned (Bhattacharyya 2011). Active riverbeds, high-flow channels and floodplain wetlands should be delineated and settlements on active channel should be discouraged, both as a matter of conservation and of public health and welfare.

5.6.1 The Need for Integrating Watershed Science and Management

Policymakers working in the domain of flood and water resource management are faced with the enormously challenging task of parsing vast amounts of data and, perhaps most importantly, understanding the complex dynamics of physical, biological and social subsystems related to the river environment. To arrive at rational decisions that will affect the lives and livelihoods of millions calls for a truly outstanding ecological understanding in the most general sense of the word ecology.

The scope and scale of these problems put a premium on developing accessible decision support systems suitable to link scientific databases, modelling analyses and geographic information systems (GIS) processing. GIS is already widely used in the field of flood and river resource management (Correia et al. 1998) and more recently in studies of the impact of sea level rise due to climate change on coastal communities (Dasgupta et al. 2009). Collaborative exploration of alternate management scenarios using GIS-based frameworks can be an essential component of linking government policy and action with academic expertise and local stakeholder interests (Burrough and McDonnell 1998; Stevenson et al. 2008; Wiley et al. 2008). Some of the important matters to be addressed on the Damodar include how much water do we now have in the river catchment, how much are we using and how much will we need later (Maidment 2002). These questions were also posed by M. Gordon Wolman (Kobell 2007). Therefore, there is an urgent need to compile and update flood- and water-related data and to make them systematically available to all stakeholders by computerisation. GIS can play a key role in making the data and their analysis explicable to both managers and the public, integrating them into a

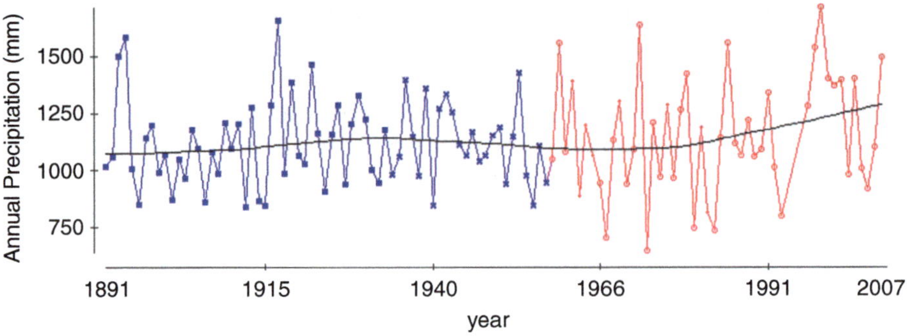

Fig. 5.9 Long-term trends in annual precipitation for combined catchments above Maithon and Panchet. Solid points (*blue*) are pre-DVC dam construction, open points (*red*) are post-dam (Data source: DVC, Maithon)

public and participatory conversation about the future. The European Union's Water Framework Directive (EP/EC 2000) provides a useful example of a framework for building the capacity of communities to conduct long-term planning at the basin-scale and manage water resources in a more deliberate and efficient manner.

Climate change potentially introduces a new and ominous feature into the functional relationship between river control practices and flood hazard, at least on the Damodar. Anticipated increases in the mean and variance of precipitation and floods in this region (Roy and Mazumdar 2005) are consistent with observed monsoon rainfall patterns in the upper Damodar catchment (Fig. 5.9) and with increases in the variability of flood flows observed post 1958. If the trajectory of the regional climate is indeed towards more extreme precipitation patterns, flood hazard will be potentially even more severe in the coming years as channel conveyance capacity continues to equilibrate to low post-dam flow and sediment regimes while extreme events nevertheless increase in magnitude. 'It's something that's been neglected, hasn't been talked about and it's something the world will have to do./.../Adaptation is going to be absolutely crucial for some societies' (Rajendra Pachauri, chairman of the Intergovernmental Panel on Climate Change, cited in Borenstein 2009). In other words, growing water consumption, hydrological, land use and climatic change acting together ensure that our rivers as well as our watersheds will face a future of rapid ecological change.

To prepare for the future, risks and options need to be effectively communicated with and effectively explored by stakeholders as well as by decision-makers at multiple levels of governments and across the Damodar Valley Corporation. Our goal should be the establishment of coherent basin-wide, integrated analytical frameworks for both hydrologic/ecological and social impact forecasting (e.g. Wiley et al. 2010). Such a framework could make use of extensive use of GIS technologies to both perform analyses and effectively communicate with the regional stakeholders. The risks and uncertainties inherent for communities living in close symbiosis with a river system are vast. Natural flood hazards and variability in geomorphic process are combined with complex social processes and larger-scale forces of climate change, regional land cover transformation and a changing political will for expanding and/or maintaining the existing engineering infrastructure. Changes in land use and climate along with hydro regime management scenarios should be quantitatively evaluated and graphically communicated, thus providing a unified basis for local and regional discussions on preparation and mitigation. The development of such a capacity would place the Damodar Valley Corporation (DVC) and its hydraulic society at the forefront of national water resource and climate change planning and policy development.

Acknowledgements We are thankful to K. K. Chakraborty (Ex Manager Reservoir Operation (MRO), DVC Maithon) and A. K. Pal (Damodar Canals No. 11 Subdivision-Rhondia) for providing valuable data. Several surveyors of RR&RD (Refugee Relief and Rehabilitation Department) at Bankura have provided data for generating land use maps. It would be a long list if we had to acknowledge within this short campus all those villagers who became friends during the field survey at various stages. We appreciate your generous help.

References

Baghel R, Nüsser M (2010) Discussing large dams in Asia after the World Commission on dams: is a political ecology approach the way forward? Water Altern 3(2):231–248

Basu AK (1989) Ganga Pather Etikatha [History of Ganges, in Bengali]. West Bengal. State Book Board, Calcutta, pp 35–36

Basu M, Bhattacharyya K (1991) Flood disaster reduction measures in the Lower Damodar, West Bengal, India, (Abstr). In: International symposium on flood disaster reduction in South-East Asia, Waseda University, Tokyo, 10–11 Nov

Batalla RJ, Gomez CM, Kondolf GM (2004) River impoundment and changes in flow regime, Ebro River basin, northeastern Spain. J Hydrol 290:117–136

Bergkamp G, McCartney M, Dugan P, McNeely J, Acreman M (2000) Dams, ecosystem functions and environmental restoration, World Commission on Dams Thematic Review – Environmental Issues II.1. Secretariat of the World Commission on Dams, Cape Town

Bhattacharyya K (1959) Bangla Deshar Nad-Nadi O Parikalpana [River planning of Bengal]. Bidyodaya library Pvt. Ltd., Calcutta, pp 69–87, also reprinted in part Utsa Manus (1990), pp 46–48

Bhattacharyya K (1998) Applied geomorphological study in a controlled tropical river – the case of the Damodar between Panchet reservoir and Falta. The University of Burdwan, Burdwan

Bhattacharyya K (1999) Floods, flood hazards and hazard reduction measures: a model – the case in the Lower Damodar River. Indian J Landsc Syst Ecol Stud 22(1):57–68

Bhattacharyya K (2002) Damodar Nad: Band Nirmaner Agge O Pare [Damodar River: in pre and post-dam period in Bengali]. Pratiti, Hooghly, West Bengal, pp 21–46

Bhattacharyya K (2011) The Lower Damodar River, India. Understanding the human role in changing fluvial environment, Advances in Asian Human-Environmental Research 3. Springer, Berlin

Bird JF (1980) Geomorphological implications of flood control measures. Aust Geogr Stud 18(2):169–183

Borenstein S (2009) Global warming may require higher dams, stilts. The Associated Press. http://www.physorg.com/news179076522.html. Accessed 3 Dec 2009

Burrough PA, McDonnell RA (1998) Principles of geographical information systems. Oxford University Press, New York

Chandra S (2003) India: flood management. Damodar River basin, Integrated flood management. Case studies. WMO/GWP Associated Programme on Flood Management, Geneva
Chien N (1985) Changes in river regimes after the construction of upstream reservoirs. Earth Surf Process Landforms 10:143–159
Chin A, Harris DL, Trice TH, Given JL (2002) Adjustment of stream channel capacity following dam closure, Yegua Creek, Texas. J Am Water Resour Assoc 38:1521–1531
Correia FN, Fordham M, Saraiva MG, Bernado F (1998) Flood hazard assessment and management: interface with the public. Water Resour Manag 12:209–227
Crutzen PJ, Stoermer EF (2000) The "Anthropocene". IGBP Newsl 41:17–18
D'Souza R (2006) Drowned and dammed: colonial capitalism and flood control in Eastern India. Oxford University Press, New Delhi
Dasgupta S, Laplante B, Meisner C, Wheeler D, Yan J (2009) The impact of sea level rise on developing countries: a comparative analysis. Clim Chang 93(3–4):379–388
Dickens CH (1853) Memorandum on the survey of the Damodar and question of the abandonment of bunds on the right bank to accompany the map received with superintending engineer South-Eastern Province's letter No. 1473 13 July, Bengal Govt. Selection 15. 1854. Bengal Military Orphan Press, Calcutta, pp 68–125
Doyle MW, Stanley EH, Harbor JM, Grant GS (2003) Dam removal in the United States: emerging needs for science and policy. EOS Trans Am Geophys Union 84(4):29–36
DVC (Damodar Valley Corporation) (1957) Report of the Lower Damodar Investigation Committee. DVC, Calcutta
DVC (Damodar Valley Corporation) (1995) Data book. DVC, Calcutta
Ehlers E (2008) Das Anthropozän: Die Erde im Zeitalter des Menschen. Wissen-schaftliche Buchgesellschaft, Darmstadt
EP/EC (European Parliament/European Council) (2000) Directive 2000/60/EC of the European Parliament and the Council of 23 October 2000 establishing a framework for Community action in the field of water policy. http://eurlex.europa.eu/LexUriServ/LexUriServ.do?uri=OJ:L:2000:327:0001:0072:EN:PDF. Accessed 30 May 2011
Gastrell JE (1863) Statistical and geographical report of the District of Bancoorah. Bengal Secretariat Office, Calcutta
GoI (Government of India) (1980) Report of the National Commission on Flood, vol 2. Ministry of Energy and Irrigation, Department of Irrigation, New Delhi
Graf WL (1999) Dam nation: a geographic census of American dams and their large-scale hydrologic impacts. Water Resour Res 3:1305–1311
Graf WL (2006) Downstream hydrologic and geomorphic effects of large dams on American Rivers. In: James LA, Marcus WA (eds) The human role in changing fluvial systems. Proceedings of the 37th Binghamton symposium in geomorphology. Elsevier, Amsterdam, pp 336–360
Grantham T, Christian-Smith J, Kondolf GM, Scheuer S (2008) A fresh perspective for managing water in California: insights from applying the European water framework directive to the Russian river, Water Resources Center contributions 208. University of California, Berkeley
Kobell R (2007) 'Reds' Wolman preaches respect for water supply. Baltimore Sun, 20 August 2007. http://www.jhu.edu/clips/2007_08/20/gulp.html. Accessed 24 Oct 2008
Kondolf GM (1997) Hungry water: effect of dams and gravel mining on river channels. Environ Manag 21(4):533–551
Kondolf GM, Batalla RJ (2006) Hydrological effects of dams and water diversions on rivers of Mediterranean-climate regions: examples from California. In: Garcia C, Batalla RJ (eds) Catchment dynamics and river processes: Mediterranean and other climate regions. Elsevier, London, pp 197–211
Kundzewicz ZW, Kaczmarek Z (2000) Coping with hydrological extremes. Water Int 25(1):66–75
Kundzewicz ZW, Schellnhuber H-J (2004) Floods in the IPCC TAR perspective. Nat Hazard 31(1):111–128
Maidment DR (ed) (2002) Arc hydro: GIS for water resources. ESRI Press, Redlands
Messerli B, Grosjean M, Hofer T, Núñez L, Pfister C (2000) From nature-dominated to human-dominated environmental changes. Quat Sci Rev 19(1–5):459–479

Molle F (2006) Planning and managing water resources at the river basin level: emergence and evolution of a concept, Comprehensive assessment research report 16. International Water Management Institute, Colombo
O'Malley LSS, Chakravarti M (1909) Bengal District Gazetteers (Howrah). Bengal Secretariat Book Depot, Calcutta
O'Malley LSS, Chakravarti M (1912) Bengal District Gazetteers (Hooghly). Bengal Secretariat Book Depot, Calcutta
Pahl-Wostl C, Craps M, Dewulf A, Mostert E, Tabara D, Taillieu T (2007) Social learning and water resources management. Ecol Soc 12(2):5. http://www.ecologyandsociety.org/vol12/iss2/art5/. Accessed 17 June 2010
Pangare G, Nielsen TK, Bhatia AM, Makin IW (2009) Innovations and advances in basin management in Asia. Discussion paper. http://www.adb.org/documents/events/2009/world-water-week/GPangare-Fpaper.pdf. Accessed 5 Jan 2010
Petts GE (1984) Impounded rivers: perspectives for ecological management. Wiley, Chichester
Richter BD, Postel S, Revenga C, Scudder T, Lehner B, Churchill A, Chow M (2010) Lost in development's shadow: the downstream human consequences of dams. Water Altern 3(2): 14–42
Roy PK, Mazumdar A (2005) Hydrologic impacts of climatic variability on water resources of Damodar river basin in India. In: Wagener T, Franks S, Gupta HV, Bøgh E, Bastidas L, Nobre C, Oliveira Galvão C (eds) Regional hydrological impacts of climatic change. Impact assessment and decision making, IAHS Red Books 295. IAHS, Wallingford, pp 147–156
Sage WM, Simms FW, Mc'Lelland J (1846) Report on the embankments of the rivers of Bengal. Military Orphan Press, Calcutta
Saha SK (1981) River basin planning as a field study: design of a course structure for practitioners. In: Saha SK, Barrow CJ (eds) River basin planning theory and practice. Wiley, Chichester, pp 9–40
Saha SK, Barrow CJ (1981) Introduction. In: Saha SK, Barrow CJ (eds) River basin planning theory and practice. Wiley, Chichester, pp 1–7
Sen SK (1962) Drainage study of lower Damodar Valley. DVC, Calcutta
Sen PK (1985a) Environmental changes and degradation in the Damodar basin: impact of development on environment. Geogr Soc India Calcutta 1:6–12
Sen PK (1985b) The genesis of flood in the lower Damodar catchment (with a case study of the September–October, 1978 flood). In: Sen PK (ed) The concepts and methods in geography. The University of Burdwan, Burdwan, pp 71–85
Sengupta DN (1951) An outline of the river problems of West Bengal. Calcutta General Printers and Publishers, Calcutta, pp 31–63
Stevenson RJ, Wiley MJ, Gage SH, Lougheed VL, Riseng CM, Bonnell P, Burton TM, Hough RA, Hyndman DW, Koches JK, Long DT, Pijanowski BC, Qi J, Steinman AD, Uzarski DG (2008) Chapter 19: Watershed science: essential, complex, multidisciplinary and collaboratory. In: Ji W (ed) Wetland and water resource modeling and assessment a watershed perspective. Taylor & Francis, London, pp 231–246
Turner BL, Clark WC, Kates RW, Richards JF, Mathews JT, Meyer WB (eds) (1990) The earth as transformed by human action. Cambridge University Press, London
UNESCO (1971a) Discharge of selected rivers of the world, vol 11. UNESCO, Paris, p 101
UNESCO (1971b) Discharge of selected rivers of the world, vol 111. UNESCO, Paris, p 59
UNESCO (1979) Discharge of selected rivers of the world, vol 111 (part 111). UNESCO, Paris, p 59
UNESCO (1985) Discharge of selected rivers of the world, vol 111 (part IV). UNESCO, Paris, p 77
Voorduin WL (1947) Preliminary memorandum on the unified development of the Damodar River. Central Technical Power Board, Calcutta
WCD (World Commission on Dams) (2000) Dams and development. A new framework for decision making. Earthscan, London

White GF (ed) (1977) Environmental effects of complex river development. Papers from a symposium of the International Geographical Union. Westview Press, Boulder

Wiley MJ, Pijanowski BC, Stevenson RJ, Seelbach PW, Richards PL, Riseng C, Hyndman DW, Koches JM (2008) Chapter 20: Integrated modeling of the Muskegon River: tools for ecological risk assessment in a Great Lakes Watershed. In: Ji W (ed) Wetland and water resource modeling and assessment: watershed perspective. Taylor & Francis, London

Wiley MJ, Hyndman DW, Pijanowski BC, Kendall AD, Riseng C, Rutherford ES, Cheng ST, Carlson ML, Tyler JA, Stevenson RJ, Steen PJ, Richards PL, Seelbach PW, Koches JM, Rediske RR (2010) A multi-modeling approach to evaluating climate and land use change impacts in a Great Lakes river basin. Hydrobiologia 657(1):243–262

Willcocks W (1930) The overflow irrigation of Bengal. Lectures on the ancient system of irrigation in Bengal and its application to modern problems. Calcutta University, Calcutta

Williams GP, Wolman MG (1984) Downstream effect of dams on alluvial rivers. U.S. Geological Survey Professional Paper 1286

Zalasiewicz J, Williams M, Smith A, Barry TL, Coe AL, Bown PR, Brenchley P, Cantrill D, Gale A, Gibbard P, Gregory FJ, Hounslow MW, Kerr AC, Pearson P, Knox R, Powell J, Waters C, Marshall J, Oates M, Rawson P, Stone P (2008) Are we now living in the Anthropocene? GSA Today 18(2):4–8

Chapter 6
Environment in an Emerging Economy: The Case of Environmental Impact Assessment Follow-Up in India

Nirmalya Choudhury

Let us all accept the reality that there is undoubtedly a trade-off between growth and environment. In arriving at decisions to untangle the trade-off, three options present themselves – 'yes', 'yes but' and 'no'. …The 'but' often takes the form of conditions that must be adhered to before, during the construction, and after the launch of the project. I believe that in laying down these conditions, we must strive for three things: First, the conditions must be objective and measurable, so that it is clear what is to be done and whether it has been complied with…

(Jairam Ramesh, Ex-Environment and Forest Minister, India (2010, p. 3)).

Abstract This chapter provides an analysis of Environmental Impact Assessment follow-up in India. It argues that its use and particularly the application of its monitoring-evaluation-management component show a consistent decline. The chapter also argues that the judiciary is one of the key drivers of EIA follow-up in the country, apart from the regulatory agency and the community. The chapter identifies the instrumental role of environmental clearance in India, the market-friendly macro-political environment and the lack of institutional strengthening of regulatory institutions as the major causes of this decline.

Keywords Environmental Impact Assessment • Environmental Impact Assessment follow-up • Compliance monitoring • India • Hydropower

N. Choudhury (✉)
Senior Research Associate, Council on Energy Environment and Water (CEEW), Thapar House 124, Janpath, New Delhi 110001, India
e-mail: nirmalya1511@gmail.com

M. Nüsser (ed.), *Large Dams in Asia*, Advances in Asian Human-Environmental Research, DOI 10.1007/978-94-007-2798-4_6,
© Springer Science+Business Media Dordrecht 2014

This chapter explores the component of *yes but* by analysing the evolution of Environmental Impact Assessment (henceforth EIA) follow-up in India. The chapter begins by defining EIA follow-up, then identifies the rationale behind the EIA follow-up exercise and explains the various *drivers* and *contextual variables* that influence its nature and outcomes. Subsequently, the chapter analyses the evolution and coverage of EIA follow-up over the last three decades of environmental governance in India. At the macro-level, the policies and activities of the Ministry of Environment and Forest (MoEF) are analysed. At the project level, the chapter analyses EIA follow-up in various hydroelectric projects.

6.1 Methodology

This chapter uses a qualitative research methodology which includes a review of various governmental and non-governmental documents: legal documents, policy documents of the MoEF, the Government of India and published and unpublished documents by civil society organisations. Semi-structured interviews were conducted with key informants identified through a purposive-stratified sampling and snowballing approach (Patton 2002). A systematic review of the Annual Reports of the MoEF from 1987 to 2011 (GoI/MoEF 1987–2011) was undertaken to review the EIA follow-up activities of the ministry over the years. Primary data was collected through case study research on the Allain Duhangan Hydro Power Project using participatory observation, semi-structured interviews with key informants and a review of various project-related documents. Case study research on the Malana Hydroelectric Project, and on a cluster of hydroelectric projects in Himachal Pradesh, is based on reviews of the interim orders by the Himachal Pradesh High Court.

6.2 Environmental Impact Assessment Follow-Up

Environmental Impact Assessment, as a planning tool, was conceptualised as an outcome of a rational decision-making process in which analysis of more information during the planning stage was expected to reduce the uncertainties and adverse consequences in the later construction and operation stages of a project (Arts and Morrison-Saunders 2006; Cashmore 2004). But as Arts and Nooteboom (1999) argue, in spite of information-intensive planning, there are always uncertainties and gaps in knowledge, and there is a great possibility that actual impacts can be different from what was predicted. This *implementation gap* (Arts et al. 2001), the gap between predicted impact and actual impact, is not just a function of missing information and knowledge but also results from gaps appearing during the implementation of the designed mitigation measures. From the point of view of environmental and social sustainability of projects, *actual impacts* are more important than *expected* impacts. In such a context, monitoring of the various planned mitigation measures and their effects on the environment, followed by evaluation and environmental auditing, is required to undertake management responses to minimise adverse environmental impacts. Hence, there is a need for an EIA follow-up.

EIA follow-up is a process which includes a range of activities, designed during the environmental clearance stage of a project and implemented during the construction, operation and decommissioning phases of a project (Jha-Thakur et al. 2009). These activities include *monitoring*, *evaluation*, *management* and *communication* (Arts et al. 2001). EIA follow-up in general, and monitoring and evaluation in particular, is undertaken for better *control*, *management*, *knowledge* and *acceptance-legitimisation* of a project (Arts and Nooteboom 1999).

The term *control* includes both the activities of checking whether the designed mitigation measures are implemented in accordance with the plan and checking if the mitigation measures are able to minimise the adverse impacts from the project through adjustments in legal-administrative decisions or enforcement actions. In this regard, it is important to undertake monitoring of both compliance and effects. *Compliance monitoring* focuses on the question of whether the project complies with the designed response strategies mentioned in the Environment Impact Assessment, the Environmental Management Plan and the Environmental Clearance Permit. *Effects monitoring* reviews the impacts resulting from a particular project and assesses whether the designed mitigation measures indeed mitigate the adverse environmental and social impacts (Arts and Nooteboom 1999). In an ideal scenario, compliance monitoring should be enough to mitigate the adverse impacts of a project, but where causal effects of adverse impacts are uncertain and there are knowledge gaps (ibid., Arts and Morrison-Saunders 2006), effects monitoring is at least equally – if not more – important than compliance monitoring.

Management or *adaptive management* (Morrison-Saunders et al. 2001), an important rationale behind an EIA follow-up, deals with taking necessary actions, in case the monitoring and evaluation activities indicate that the mitigation measures fall short of having the desired impact. Arts and Meijer (2006) argue that in cases where there is limited possibility of taking management response through adapting the mitigation activities, undertaking an EIA follow-up is of little use.

EIA follow-up also helps closing existing knowledge gaps. In the short and medium terms, the EIA follow-up process strengthens the planning and operation of ongoing projects through monitoring and evaluation of impacts. This is often defined as *single-loop learning*. In the long term, it not only strengthens the environmental governance of a particular project but also fortifies the effectiveness of the overall EIA system. EIA follow-up not only ensures learning at the project level but also allows critical reflection on the EIA system in general. The monitoring and evaluation of real-time environmental data strengthens the understanding of *cause-effect relationships* and leads to better prediction of the environmental impacts in the future. This is often defined as *double-loop learning* (Marshall et al. 2005; Morrison-Saunders and Arts 2005; Marshall 2005; Arts and Nooteboom 1999).

Last but not least, an EIA follow-up is also undertaken by the regulatory agency and the project developer in order to enhance trust. Through various forms of *communication*, from public communication over websites to public participation in monitoring committees, an EIA follow-up ensures that the communities remain aware of the activities undertaken by a project (Arts et al. 2001). This process legitimises the environmental clearance process in the eyes of the community and justifies the continuation of ongoing operation by providing a sort of a social licence.

6.2.1 What Drives EIA Follow-Up?

EIA follow-up is driven by a combination of three drivers: the project proponent, the regulatory agency and the local community. The reason to conduct an EIA follow-up can be different for different drivers.

The project proponent-initiated EIA follow-up – also known as first-party EIA follow-up – is independently initiated by the project developer, even if he is not legally bound to do so. This form of follow-up is proactive and anticipatory (Morrison-Saunders et al. 2001).

An EIA follow-up could also be driven by a regulatory authority. In this case the permission to go ahead with a project is contingent upon its environmental clearance. This includes the preparation of an Environmental Impact Statement, an Environmental Management Plan and a definite plan of monitoring and reporting of the mitigation measures requested by the regulatory authority. Thus, the regulator-driven EIA follow-up takes place as an act of compliance to the legal statutes of the country and might be the result of sustained public pressure against the adverse impacts of a project. There are basically two types of institutional arrangements for regulator-driven EIA follow-ups: either a team is employed by the regulator to monitor and manage the EIA follow-up (*environmental team model*) often by setting up a new project office or an independent agency is contracted to check the EIA follow-up (*independent checker model*) (Morrison-Saunders et al. 2001, 2003).

The third important driver of an EIA follow-up is the *community*. This can take many forms, from a local community immediately affected by a project to a consortium of international pressure groups criticising large infrastructure projects (ibid.). It has been argued that EIA follow-ups driven by community mobilisation are mostly *reactive*. It takes place after an adverse impact has already occurred and the local community feels that the project has negative social and biophysical impacts which remain unmitigated and thus affect the populace. In such cases, an EIA follow-up would result in the reduction of the insecurity among the community regarding the adverse impacts of the project. Additionally, it would enhance their knowledge, and most importantly, it would provide them an arena to deliberate with the project managers and the regulators on issues related to environmental impact mitigation and compliance (Morrison-Saunders et al. 2001; Arts et al. 2001).

An EIA follow-up in a country is not necessarily the outcome of just one of these drivers. For instance, an EIA follow-up could be initiated by a project developer – or more importantly by a project sponsor, as often seen in developing countries – but in the absence of any legal backdrop, it might be the regulatory authority that wishes to bring the necessary legal changes. Similarly, it could be community pressure which coerces and persuades the project proponent or the regulatory agency to initiate an EIA follow-up.

6.2.2 What Affects EIA Follow-Up?

The design and outcome of an EIA follow-up is also found to be affected by a number of contextual conditions, i.e. the external environment within which it operates. These conditions remain exogenous to the system and define the limits of the different drivers for having an impact on the process and outcome of an EIA follow-up. When an EIA follow-up is analysed over a long period of time, some of these contextual variables may not remain constant.

One of the important contextual variables affecting the outcome of the EIA follow-up is the legal and administrative *enactment of statutory notifications,* in the form of resolutions, notifications or laws. Such enactments not only provide a statutory backup for follow-up activities but also standardise them, rather than keeping them voluntary or ad hoc. In addition they empower the regulatory authorities to take proper sanction measures in cases of non-compliance (Morrison-Saunders et al. 2003).

Another important contextual variable that affects the process and outcome of an EIA follow-up is the very *nature of the project.* The design of an EIA follow-up is especially important for projects which involve large capital investments and a high degree of uncertainty and unfamiliarity (both in terms of impacts and in terms of the suggested mitigation measures). They are also important for projects located in socially or ecologically sensitive areas, where the risk of adverse impacts is especially high.

EIA follow-ups also depend on the *approaches and techniques* undertaken to implement them. In this regard, international experience shows that a proactive (early) well-designed EIA follow-up is always better than a reactive one, which takes place when the impacts have already manifested and the project is subject to criticism from diverse stakeholders (ibid.). In the latter case, an EIA follow-up, rather than being a management tool for sustainability, becomes a tool for defending the actions and activities already taken. Thus, an early and well-designed EIA follow-up has been proclaimed as 'winning half the battle' (Jha-Thakur et al. 2009; Arts and Meijer 2006).

The effectiveness of an EIA follow-up also depends on the *resources and capacities* of the different stakeholders involved in its planning and implementation. In this regard, a higher degree of community involvement enhances the effectiveness of the EIA follow-up exercise. At a minimum, this should include some form of information provision. In better cases it includes community participation in the form of multi-stakeholder monitoring groups (Arts et al. 2001; Morrison-Saunders et al. 2001, 2003).

6.3 Trends in EIA Follow-Up in India

6.3.1 The Pre-1994 EIA Notification Era

The process of undertaking an EIA as a part of the environmental clearance requirement dates back to the late 1970s, when it was first applied to river valley and hydropower projects. The environmental clearance was in most cases then – and is still now – conditional. Sometimes, these conditions were the outcome of an EIA where the environmental clearance of a project was subject to compliance with a set of conditions, mentioned in the Environmental Management Plan and the clearance permit. For others, predominantly river valley and hydropower projects like the controversial Sardar Sarovar Project, the environmental clearance was awarded under the condition that further studies to determine the actual impact of the project would be carried out, and that mitigation plans along with the construction of the project would be designed and implemented. Such cases intuitively required the monitoring, evaluation and management of the social and environmental impacts of the projects. Thus, quite expectedly, the Annual Reports in the late 1980s and early 1990s had separate components on the Sardar Sarovar Project (and also Tehri Hydropower Project which was also becoming contentious). These showed that the Ministry of Environment and Forest gave a lot of importance to monitoring the status of compliance with mitigation conditions (GoI/MoEF 1989, 1990, 1991, 1992, 1993). To a certain extent, the Sardar Sarovar Project and Tehri Hydro-Electric Project ensured that compliance monitoring remained a distinct activity of the Ministry of Environment and Forest.

As early as 1986–1987, the Annual Report of the MoEF mentioned the importance of 'effective implementation of the suggested mitigative measures' and suggested strengthening this process institutionally through the formation of 'monitoring committees' which would ensure that the mitigation measures were undertaken prior to the onset of the actual operations (GoI/MoEF 1987, p. 28). However, due to the absence of statutory obligation, the lack of independent administrative strength to undertake monitoring during the post-clearance stage and the absence of punitive clauses in cases of non-compliance, the recommendations of the ministry depended heavily on the discretion of the project developer. Thus, with respect to monitoring and implementing mitigation measures, the EIA follow-up was driven by administrative requirements and by the MoEF. The actual implementation of the same instead depended on the project developer.

The observed developments in the late 1980s were also an outcome of a couple of contextual variables. First, an umbrella legal instrument titled the *Environmental Protection Act 1986* – and the *Forest Conservation Act 1980*, few years earlier – legally empowered the MoEF to take action to protect the environment. More importantly, there was public pressure and agitation against the adverse social and environmental impacts of river valley and hydropower projects, many of which included the construction of dams. Thousands of these have been built since independence[1]. Their adverse social and environmental impacts were rarely captured on the conventional planning radar, which was still focused on conventional cost-benefit-type analyses.

In the late 1980s, the MoEF set up *regional offices* in different parts of the country. Their task was predominantly to monitor compliance with the mitigation measures, as prepared by the project developers in their Environmental Management Plans (GoI/MoEF 1989). This institutional set-up was then followed by an operational plan pertaining to the reporting and monitoring of the Environmental Management Plans. These were six-monthly progress reports to be prepared by the project proponents in order to report on the progress of compliance with the activities mentioned in the Environmental Management Plans, followed by a field review by the personnel from the regional offices of the MoEF. It also included an analysis of the implementation process and a communication of the findings between the project proponent and the MoEF's regional offices to take remedial measures, if these were required. The MoEF also designed sanction measures to deal with non-compliance. The first was to report to appropriate state government authorities. The second was to review the issued environmental clearance if the project developer had changed the scope of the project after project clearance without informing the environment ministry (GoI/MoEF 1991, 1992).

Through these institutional mechanisms, the environment ministry ensured that different components of the EIA follow-up, like monitoring and midcourse correction through adaptive management by the proponents, were incorporated into the overall environmental governance process. In spite of public pressure and agitation being a driver of EIA follow-up, the EIA process in general and EIA follow-up in particular were not open to public involvement. Monitoring undertaken during an EIA follow-up, as well as subsequent communications and actions, was largely an interdepartmental affair. Thus, the developments during this period resembled regulator-driven EIA follow-up. With setting up of new project offices, in this case regional offices, the institutional structure for EIA follow-up reflected *environmental team models*.

Since the early 1990s, apart from issuing environmental clearance, the monitoring of projects which had received environmental clearance was an important activity of the MoEF. Thus, until 1994, the year when the first EIA Notification was passed, the Annual Reports always mentioned the total number of projects under different sectors that were monitored in the given year along with their annual project clearance status (GoI/MoEF 1992, 1993, 1994b).

While the Annual Reports addressed EIA follow-up in terms of the number of projects monitored, there was no information on the *learning* that resulted from the monitoring-evaluation-management process performed by the MoEF. Among the various rationales for undertaking EIA follow-up, the learning component seems to

[1]To date, India has built around 5,000 'large dams' (according to the definition of International Commission of Large Dams).

fade out. It can be inferred that, although the scope of EIA follow-up was defined quite broadly – and ambitiously – by MoEF, its implementation was reduced to just another bureaucratic activity having little or no learning value.

6.3.2 The Post-1994 EIA Notification Era

EIA Notification 1994 (GoI/MoEF 1994b) converted environmental clearance and EIA from *administrative* to a *statutory requirement*. However, the notification did not change the ongoing EIA format that had evolved as an administrative requirement in the early 1990s. EIA Notification 1994 made environmental clearance legally mandatory, and separate roles were demarcated for different stakeholders. The project proponent was responsible for undertaking the EIA. It had to include Environmental Impact Statements, as well as suggestions regarding the mitigation measures to be undertaken during the course of the project. The MoEF would act as a competent regulatory authority which would issue environmental clearance. Furthermore, a multidisciplinary expert committee called an Expert Appraisal Committee (henceforth EAC) was set up to evaluate the EIA. Its objective was to recommend either an unconditional or a conditional environmental clearance, or a rejection of environmental clearance to the MoEF. The regional offices of the environment ministry were entrusted with the activity of compliance monitoring.

EIA Notification 1994 also empowered the EAC to undertake site visits after environmental clearance was issued, in order to verify whether the conditions of the issued environmental clearance were being complied with during the project's construction-operation phase. But in spite of this statutory power, Roy (2008) notes that at various stages, compliance was often delayed by the project developer and that this largely went unnoticed by the EAC. This happened because the EAC hardly ever used the provision made in the EIA Notification 1994 to monitor compliance. The field visits that *were* made – if indeed any at all took place – happened only during or prior to the process of awarding environmental clearance, and not thereafter.

The disparity between the legal statutes of EIA Notification 1994 and the reality of practice on the ground is reflected in the various Annual Reports of MoEF after 1994. It shows how the importance of activities like monitoring, evaluation and adaptive management in the form of midterm correction became diluted over time. Monitoring, the first step towards EIA follow-up, which was a separate activity component mentioned in the MoEF's Annual Reports until 1994, is found to be particularly missing in the years after the 1994 Notification (GoI/MoEF 1996, 1997, 1998). This shows – quite contrary to what could have been expected from EIA follow-up experiences in other countries – that in spite of a statutory backup, the importance of EIA follow-up in general and compliance monitoring and effects monitoring specifically, at least from an administrative perspective, dwindled over the years. This happens to the extent that they find no place in the Annual Reports, which are documents that are supposed to list the important administrative and regulatory activities undertaken during these years.

Since 1999, the Annual Reports of the MoEF restrict the information on EIA follow-up to the gross number of projects monitored each year. They provide no information on whether the monitoring included desk reviews of the six-monthly compliance reports handed in by the project proponent, or whether they were an outcome of the field review undertaken by the regional offices. The Annual Reports of every single year since 1987 comprise *no* information on *effects-monitoring*. Also, the Annual Reports do not mention any results of the mitigation activities undertaken by the project proponents, nor whether such information resulted in project modification or system-wide modification (the single- and double-loop learning processes mentioned earlier). However, by the early 2000s, some learning from the *compliance monitoring* seemed to take place, as the ministry identified it as one of the weak areas within the overall process of environmental governance. This is reflected in the various Annual Reports between 2001 and 2005. In these, the MoEF highlighted and repeated the major areas of non-compliance year after year, without changing even the sequence in which these areas were mentioned. Additionally, no information on the activities taken up by the MoEF to address those weaknesses, nor their outcomes, were mentioned in these Annual Reports (GoI/MoEF 2001, 2002, 2003, 2004, 2005).

The environmental clearance and, more importantly, the EIA and its subsequent follow-up during the design stage were contingent on three things: the accurate portrayal of an environmental baseline, the scope of the project and the likely socio-environmental impacts. EIA Notification 1994 made provisions for strong sanction measures (including outright rejection of environmental clearance) in cases where proponents were found to misrepresent the facts on their Environmental Impact Statements (GoI/MoEF 1994b). But despite the legal provisions, its implementation lacked efficacy: changes in the scope of the project after environmental clearance, misrepresentation of the environmental baseline conditions, missing information and hastily added information were observed (Roy 2008; Jha-Thakur et al. 2009). Under such circumstances, one would have expected the implementation of a stronger EIA follow-up through regular compliance monitoring by the MoEF in order to *control* the adverse impacts of the project, not to mention fulfilling the *legitimising* and *learning* requirements. On the contrary, the administrative importance of compliance monitoring declined. Thus, the environmental clearance process in general and monitoring of environmental clearances in particular were found to be the *Achilles' heel* of the overall environmental governance process. Many perceived it as a hindrance in the path to rapid economic growth. The MoEF was criticised from all sides, including from civil society, academia and the business community. It was clear that the overall environmental clearance process had to be revamped (Nandimath 2009; Roy 2008; Paliwal 2006).

6.3.3 The EIA Notification of 2006

EIA Notification 2006 (GoI/MoEF 2006b) standardised the overall environmental clearance process by including screening, scoping, public consultation and appraisal as the stages of the environmental clearance process. Some of the provisions of EIA Notification 2006 were commendable though leaving sufficient room for debate on each of them. These included video recording of public hearings as a part of public consultation exercise, decentralisation of the environmental clearance process for some projects from MoEF to state-level institution and institutionalisation of screening and scoping in the environmental clearance process.

However, the new notification diluted the sanction provisions of EIA Notification 1994 in cases of submission of faulty environmental baseline information or understatement of the likely impacts. In the EIA Notification of 2006, there are provisions for sanctions, but only if it is proven that the false information was submitted 'deliberately'. Even in that case, the project is only 'liable' to rejection and not rejected automatically (GoI/MoEF 2006b, p. 8). EIA Notification 2006, instead of addressing the implementation weaknesses of the otherwise strongly worded EIA Notification 1994, diluted the sanction measures, further weakening the already frail EIA follow-up.

EIA Notification 2006 did not have a positive impact on the various components of the EIA follow-up, especially in comparison to the 1994 EIA Notification. While the EIA Notification 1994 had provisions for the EAC to make site visits to a project both during and after the environmental clearance process, the 2006 Notification restricted the site visits to the time during the environmental clearance. It is silent on site visits once the environmental clearance is issued (GoI/MoEF 2006b). This implies that the project developers would send a six-monthly compliance report to the MoEF, which would then be subjected to desk reviews by the regional offices. This dilutes the effectiveness of the EIA follow-up, whose efficacy rests on filling the information gap, uncovering uncertainties, real-time monitoring and adaptive managerial action. It became increasingly unlikely that desk reviews would yield any meaningful information on compliance monitoring (and more importantly effects monitoring). Unfortunately, while EIA Notification 2006 does address some of the procedural constraints of the EIA process, the notification fell short in addressing the even more important post-clearance activities. In sum, the broad mandate of the EIA follow-up that the MoEF had articulated in the early 1990s had become completely diluted by the end of the first decade of the new millennium.

The Annual Reports of 2007 and 2008 define the scope of compliance monitoring of environmental clearance in terms of the number of projects monitored and suggestions given to project proponents based on a *review* of the six-monthly compliance report. The Annual Reports – quite expectedly – provide no information on the number of projects on which the MoEF undertook field reviews. In addition, they contain no information neither on the learning from the monitoring exercise nor on the number of projects that were modified or sanctioned on grounds of non-compliance with the Environmental Management Plan (GoI/MoEF 2007, 2008). The absence of all this information implies that, even if monitoring *were* taking

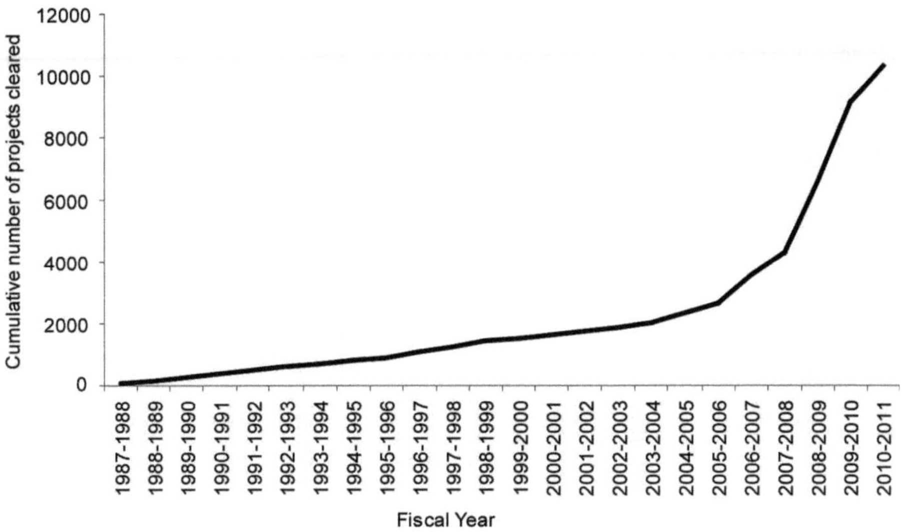

Fig. 6.1 Cumulative number of projects that received environmental clearance (Data source: GoI/MoEF 1987, 1988, 1989, 1990, 1991, 1992, 1993, 1994a, b, 1995, 1996, 1997, 1998, 1999, 2000, 2001, 2002, 2003, 2004, 2005, 2006a, 2007, 2008, 2009a, 2010a, 2011)

place, the *controlling* and more importantly *learning* through monitoring would still be missing. The Annual Report of 2009 brings a couple of virtues of the monitoring of environmental clearances back onto the agenda. But on the actual activity front, the scope of monitoring is again restricted to citing the total number of monitoring reports received and analysed. However, in recent years it could be observed that the MoEF is trying to strengthen the environmental-compliance process through necessary actions which are to be initiated for projects where non-compliance is noticed (GoI/MoEF 2009b) and through preparation of a procedure to deal with projects which show non-compliance. This includes various stages of clauses for penalties, ranging from issuing a show-cause notice to closing down a project (GoI/MoEF 2010b). Thus, even after more than three decades of EIA practice in India, the regulator-driven EIA follow-up model is still struggling to fulfil the *controlling* rationale of EIA follow-up – leave alone the *learning* rationale.

6.3.4 Trend Analysis of Coverage of EIA Follow-Up in India: 1987–2010

The use of the EIA process has increased over the years from being an administrative requirement only used for multipurpose river valley and hydropower projects in the early 1980s to a statutory requirement covering around 30 different types of projects. This has resulted in an exponential increase in the number of projects that have obtained environmental clearance over the last three decades (see Fig. 6.1).

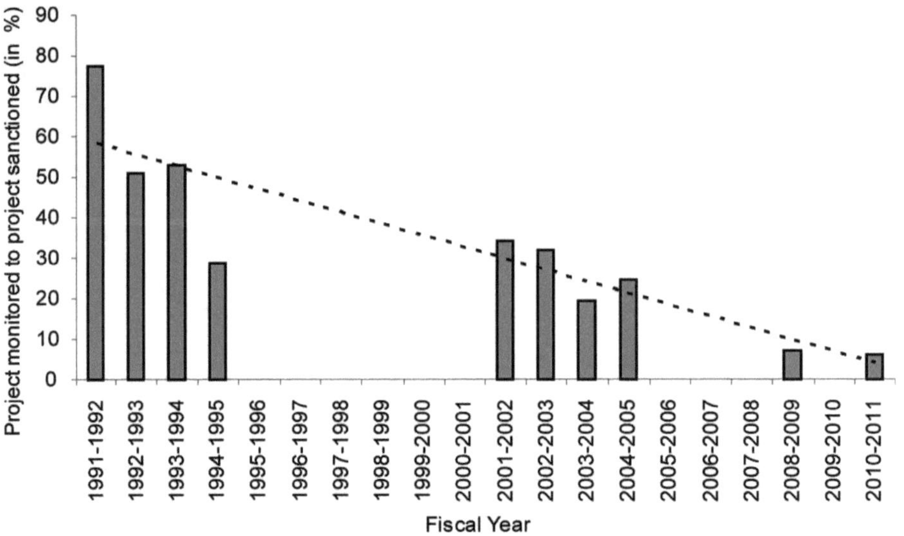

Fig. 6.2 Trend analysis of EIA coverage in India (Data source: GoI/MoEF 1992, 1993, 1994a, b, 1995, 2002, 2003, 2004, 2005, 2009a, 2011)

However, while the coverage of EIA increased consistently (in fact, exponentially in recent years), the number of projects which were subjected to compliance monitoring decreased. Thus, the coverage of EIA follow-up consistently declined over the years. The coverage of EIA follow-up (or more specifically, compliance monitoring) is measured by the ratio between the number of projects monitored in a particular year and the total number of projects that have received environmental clearance prior to the year under analysis.

The information on monitoring of environmental clearances in 10 of the last 25 years shows that the coverage of EIA follow-up over the entire period is around 36 % (with a high variation). The trend analysis (see Fig. 6.2) shows a consistently declining trend of EIA follow-up over the years – from 77 % in 1991–1992 to a mere 7 % in 2008–2009. The results confirm that over the last two and a half decades, while the coverage of EIA has increased, compliance monitoring has shown a consistent downward trend.

6.4 EIA Follow-Up in Practice: Three Case Studies

The previous section analysed the evolution of EIA follow-up, at the macro-level, both in terms of policy development and activities for compliance monitoring. This section looks closely at the project level – where the actual drama of EIA follow-up is played out. In the process, it analyses the role of different *drivers* and *contextual variables* in influencing the nature and outcome of the follow-up.

6.4.1 Allain Duhangan Hydro Power Limited

Allain Duhangan Hydro Power Limited (henceforth ADHPL) is a 192 MW run-of-the-river hydropower project in the state of Himachal Pradesh, India. It is a private-sector-owned project designed to function as a merchant power plant which will sell the bulk of its power at spot prices during peak hours. It has a high level of presence of international actors, including the project developer (a Norwegian power company) and the project financier (International Finance Corporation, henceforth IFC). Both have superior standards with higher social and environmental safeguards than the Government of India's requirements and standards. This case study sets out to verify whether the presence of international project developers and financers results in a better EIA follow-up design and execution.

During the process of environmental clearance of the project, a number of civil society associations at the national, regional and local levels got involved. They conducted awareness-raising programmes in the project area and ensured through various acts of persuasion and coercion that the IFC standards, which were higher than the national standards, were complied with during project clearance. Thus, this project was subject of a number of interventions. The EIA report was redone and the civil society had access to technical and human resources for preparing the public for the hearing. In addition, independent regulators, who were chosen by the civil society organisations, regulated the execution of the public hearing. None of these is representative of the general environmental clearance process followed by most of the projects in India. It helped to pacify the concerns of a significant proportion of the local populace. When the project finally got clearance from the IFC and construction operations began, local people also got some small contracts. As a result, a significant proportion of the local populace were no longer antagonistic towards the project. However, some continued to protest on the grounds that public hearing was undertaken on the basis of an incomplete EIA report and demanded a new public hearing (Choudhury 2010; IR 2009; IFC/CAO 2004[2]).

During the construction phase of the project, there were allegations by the Himachal Pradesh Environment and Forest Department against the project authority of repeated violations of the forest and the environmental clearances. Subsequently, a committee was formed by the Government of Himachal Pradesh to look into the matter. The committee approved that in spite of repeated notices from the state government, there were violations of forest clearance. These comprised illicit felling or damaging of trees, dumping of waste (especially material excavated during tunnelling activities) in the river and diversion of forest land beyond the permitted

[2]A part of the information here is based on interviews with South Asia Network of Dams Rivers and People (SANDRP), Kalpvriksh and Dhomiya Ganga Sangharsh Samiti, conducted in June 2009.

area without the necessary approval from the MoEF.[3] Certainly, the EIA follow-up action – if at all taken – was not able to mitigate adverse environmental impacts.

The section of the local populace who was still opposing the project clearance filed a lawsuit based on the findings of the committee and demanded to stop the work on the project. As a result of the lawsuit, the Himachal Pradesh High Court (henceforth the court) issued a Stoppage-of-Work Order on the project. The project developer appealed to the court to lift the order, and the court, in response to this appeal, appointed an expert committee comprised of district administration and senior functionaries as well as members of the Forest Department, Public Works Department, State Electricity Board, Department of Environment and Pollution Control Board. This committee was asked to undertake a site visit within 3 days and to submit a report to the court within a week. The committee was empowered by the court to remove the Stoppage-of-Work Order *if* they found that the project developer was complying with the mitigation measures and was taking actions to minimise the adverse impacts. The committee, based on the site visit, partially removed the Stoppage-of-Work Order.[4]

A second lawsuit was filed by those people who were opposing the environmental clearance of the project on the grounds that the EIA report underrepresented the environmental baseline and the Environmental Management Plan. This lawsuit was on the grounds that the EIA report and the Environmental Management Plan, which the project developer was supposed to follow-up in the post-clearance stage, did not correctly estimate the impact of the water diversion for power generation on the lives and livelihood of a downstream village. They argued that the minimum downstream flow mentioned in the EIA report was not sufficient for the livelihood requirement of the village. In response, the court appointed a number of expert committees to evaluate the water requirements for various purposes[5] of this village. As of June 2009, the matter was under judicial review.

6.4.2 Malana Hydroelectric Power Project

On 24 November 2009 *The Tribune* – one of the newspapers in Himachal Pradesh – reported that in one of the state's hydropower projects, the 100 MW Malana II project, blatant violations of the environmental and the forest clearances

[3]This is based on the review of various legal documents, which included a number of signed petitions filed by the people affected by the project and the project developer, accessed from the website of Himachal Pradesh High Court.

[4]This is based on the review of various legal documents, which included number of writ petitions filed by the people affected by the project and the project developer, accessed from the website of Himachal Pradesh High Court.

[5]This is based on interviews with some members of the community and a review of the legal documents submitted with Himachal Pradesh High Court.

during construction had occurred (Chauhan 2009). Taking notice of this news, the Himachal Pradesh High Court directed the *District Magistrate*, the highest administrative decision-maker of the Himachal Pradesh Government within the project area, to visit the project site and evaluate if the project complied with the statutory environmental and forest clearance. The report of the District Magistrate was not acceptable to the court, because it was too vague. Thus, the court constituted a two-person committee to undertake the auditing activity (HCHP 2010c).

Based on the report of this two-person committee, which discovered violations of the environmental and forest clearances, the court came down heavily on the Government of Himachal Pradesh with the scathing remark that the government failed to perform the role of 'guardians of public property' and acted 'as if they are agents of the project proponents' (ibid., p. 3). The court noted that there was 'irreversible damage' (ibid., p. 3) to the environment and criticised the Forest Department for failing to perform its duties. Then the court issued a Stoppage-of-Work Order on the project. Subsequently the court formed a one-person committee (henceforth Sharma Committee) to offer an expert opinion on the measures reported to be taken by the project proponent to minimise the adverse environmental impacts. The Sharma Committee was endowed with the power not just to evaluate or audit the environment and the forest compliance but also to suggest remedial measures to minimise the environmental impact, in order to ensure that the project henceforth would not have an adverse environmental impact (ibid.).

Based on the condition that the recommendation of the Sharma Committee would be strictly complied with by the project developer, and on reference from the highest legal representative of the Government of Himachal Pradesh, the court lifted the Stoppage-of-Work Order. Further, to ensure the implementation of a strong monitoring system, which would minimise the adverse environmental impact of future project activities, the court suggested to constitute another three-person monitoring committee. This would include the highest administrative official of the state government, the highest official from the Environment/Forest Department and the one-member Sharma Committee. The three-person monitoring committee was asked to submit a performance report of the project to the court after a period of 4 months.

In spite of this order, the three-person monitoring committee did not find the time to undertake a field review of the monitoring, evaluation and management activities – if indeed any were undertaken – by the project proponent. The court again came down heavily on the Himachal Pradesh Government, this time because of showing laxity in undertaking field monitoring and auditing despite directions from the court. The three-person monitoring committee finally visited the project and submitted a report to the court. However, the court was again not satisfied with the report and appointed the *Amicus Curiae* to visit the project area and submit another report.

The *Amicus Curiae* reported that the project proponent was complying with the conditions in certain cases, but that in others the recommendation was difficult to comply with, given the technical constraints of the project. Judging by the various photographs collected from the *Amicus Curiae's* report, the court observed that there had been serious environmental impacts, which would not have occurred without the connivance of the Forest Department or a serious laxity in the performance of the Forest Department and State Pollution Control Board, who were supposed to have been monitoring the mitigation activities and their effects on the environment (HCHP 2010b).

Subsequently, the court formed another two-person committee consisting of the one-member Sharma Committee and a technical officer from the project proponent, to undertake a joint inspection of the site. This committee was instructed to evaluate the effects of different mitigation measures on the environment and to suggest appropriate management actions for the project. This committee was asked to submit a copy of their recommendations to the court. The court rebuked the Forest Department for being lax in their task of protecting the environment. It also impressed upon the Forest Department to undertake stringent effects monitoring and not to restrict itself to mitigation of compliance monitoring or receiving compensation in case of violations. In the event of violations, the Forest Department was asked to take strong sanction measures, like stoppage of work and enforcing financial compensation much higher than the amount that is mandated statutorily (HPHC 2010a).

6.4.3 A Cluster of Hydroelectric Projects in Himachal Pradesh

On 19 November 2009, the newspaper *Indian Express* reported that large tracts of forests were subjected to illicit felling. This was due to hundreds of hydropower projects that were being built across the length and breadth of the state. The High Court of Himachal Pradesh took *suo motu* cognisance of this and constituted a one-person committee (henceforth Shukla Committee) to establish if the projects under construction were adhering to the mitigation measures and whether the mitigation measures had the desired effects of minimising the adverse environmental impacts. The court also wanted to find out if the overall process of decision-making and planning of hydropower projects across the state cumulatively addressed the environmental safeguards, or whether hydropower planning and policy needed to be revamped (HCHP 2009).

Based on the report of the Shukla Committee, the court noted that although in many cases the projects complied with the requirements of the environmental clearances, there was still evidence of violations with regard to mitigation measures. It further concluded that a lot of the adverse impacts on the environment were taking place during the construction of hydropower projects, particularly due to road construction and the lack of proper management of the disposal of excavation material. The court also noted that the extent of violation showed that during the preparation of the EIA and the Environmental Impact Statement, either the environmental baseline or the likely impacts had been understated (HCHP 2010a).

The court heavily criticised the Himachal Pradesh Government for not warranting that its various administrative departments ensured the law to be executed – in letter and spirit – by various hydropower projects (ibid.). The legislation drafted by the state required that all hydropower projects must ensure a 15 % mandatory water flow downstream of a diversion or a storage project as an Environmental Flow Requirement (EFR). According to the Shukla Committee, all but one of the projects lacked any provisions whereby the EFR could be released independently of season, duration and mode of operation. In most occasions, the Shukla Committee found that either the projects made some provision but that the minimum flow would not ensure automatic year-round release from the project, as mandated legally, and would depend on the operations of the project proponents. Or – even worse – it found that they had yet to make any provisions even where the project was in an advanced phase of construction. More importantly, the Shukla Committee found that the various relevant departments of the Himachal Pradesh government either were in many occasions agreeing to the arguments given by the project proponent on why the project did not require the release of any environmental flows or were satisfied with the oral assurances by the project proponent that adequate provisions had already been made or would be made in future, without physically verifying the fact (ibid.).

The court, based on the findings and the recommendation of the Shukla Committee, questioned the overall hydropower policy of Himachal Pradesh, where cascades of hydropower projects were allowed on many of the rivers and their tributaries without undergoing any cumulative impact assessment (basin level studies) and without mandating a minimum distance between two consecutive cascade projects in order to prevent the threat that large stretches of rivers would only run through tunnels[6] (ibid.).

The court, *prima facie*, accepted the short-term recommendations of the Shukla Committee report. The committee on numerous occasions had gone beyond mere auditing and reporting and had suggested suitable mitigation actions to be taken up by the project proponent. On many occasions, the Shukla Committee, along with the project proponent, undertook joint Environmental Management Planning, based on both compliance and effects monitoring. The court accepted these project-specific recommendations and ensured that immediate implementation of the management recommendation would be taken up by the project proponents (ibid.).

Finally, the court ordered to circulate the report of the Shukla Committee among the relevant state government departments, because numerous policy-level suggestions and gaps were highlighted which required government responses. In addition, it postulated to upload the report on the website of the High Court so that everybody could download it. In this way, the court addressed an often forgotten issue in EIA follow-up: that communication needs to go beyond the project proponents and government departments and that the general public deserves at least to be informed about the development through *public communication* (Rowe and Frewer 2005).

---

[6]The Shukla Committee found that in a 70 km long river basin, only a 3 km stretch of the river was found to flow on the original river bed, while the other 67 km of the river was diverted through the tunnels.

6.5 EIA Follow-Up: A Case of General Decline

In a country like India, where the baseline environmental database is either weak or non-existent, making an environmental impact prediction through an EIA report and drafting an Environmental Management Plan is both difficult and contentious. The level of uncertainty and the potential for conflict are often underlined with different – and often contradictory – values and belief system. The uncertainty and contentiousness is manifested starkly when it comes to large infrastructure projects like river valley and hydropower projects. They often have a substantial and differential impact on different sections of society. The river valley and hydropower projects, like many other infrastructure projects, tend to distribute costs and benefits unequally on different sections of the populace. The beneficial impacts are manifested at the regional level, while the adverse environmental and social impacts are manifested at, and are many times restricted to, the local level. In India, the indigenous people (*scheduled tribes*) usually bear the brunt of these projects. The regulatory agency and the project proponent, in such cases, tend to suffer from a deficit of trust from the civil society and the public. In absence of proper environmental and social impact analysis, the affected public see themselves losing out from a project and having no *voice* in the decision-making process (Menon and Kohli 2008, 2009). This phenomenon was observed in the case of the Allain Duhangan Project. On the other hand, the project proponents together with various growth-oriented Ministries of the Central and State Governments criticise the environment regulatory agency (Ministry of Environment and Forest) for constraining the economic growth of the country (GoI/MoF 2009). In such a scenario with a trust deficit, an inequitable benefit-risk sharing and an incentive for certain sections of the government to devalue the EIA and environmental clearance, EIA follow-up would not only ensure proper compliance monitoring but would also enhance mutual trust. Furthermore, it would at least make the project-affected community feel secure. In doing so, the adverse impacts of the project would be taken care of by the proponent and the regulatory authority. In the long run, EIA follow-up could cut down the redundancies in the EIA process and strengthen the overall environmental governance system through what have been defined earlier as single-loop and double-loop learning processes. But, as this analysis of the reality of EIA follow-up in India shows, its importance has declined over the years. Currently, its aims are at best to perform the role of compliance monitoring. This is far from an ideal situation.

Studies of EIA follow-up in other countries identified self-motivation of the project developers as one of the triggers of EIA follow-up (Morrison-Saunders et al. 2001). This is a phenomenon that is rarely, if at all, observed in India. The case study of the Allain Duhangan Project shows that even in the presence of international actors which have superior safeguard standards, one may at best expect that better norms and standards are being followed during the project planning and clearance stage. However, this does not translate into better standards in the post-clearance stage. Jha-Thakur et al. (2009) rightly argue that these result from the perceived importance of the environment in the overall scheme of things. For any project proponent, the objective function circulates around quick construction and early operation. Therefore, EIA and EIA follow-up, rather than being seen as (adaptive) management tools, are seen more as a hurdle to be overcome or bypassed. This implies that in most occasions, the importance of EIA declines when the project gets environmental clearance and is allowed to proceed from its planning to its construction-operation phase. Environmental clearance has just an instrumental value. It is akin to a licence or a permit and is nowhere close to being a management tool. This is the *instrumental reason* behind the general decline of EIA follow-up in India.

Along with the globally recognised three drivers of the EIA follow-up, namely, the project proponent, the regulatory authority and the community, the *judiciary* can also act as the fourth driver in instituting EIA follow-up. As shown by the case studies, the judiciary is an important *driver* of EIA follow-up in India. The involvement of the judiciary took place after an adverse impact. Thus, it is more often than not a reactive rather than proactive managerial approach from the project developers. But, in countries where the regulatory institutions like the MoEF or the state environment and forest departments are weak in enforcing regulations – or just not interested in doing so as revealed in this chapter – and where avoidance or evasion of rules is ubiquitous, the judiciary can at least ensure that some remedial measures are undertaken to reduce the adverse impacts (see Sect. 6.4.1).

Environmental clearance has been subjected to different pull and push factors from diverse actors with often diverse and contradictory sets of interests. The civil society continued to argue for developmental activities which had minimum environmental impact. Environmental Impact Assessment, Environmental Management Plans and EIA follow-up opened a new arena through which large water infrastructure projects could be encountered. However, the prevailing macroeconomic scenario in the country had changed since the 1990s. Under the *liberalisation* policies, the focus of the federal and various state governments has been on improving the economic parameters in order to attract private capital. Thus, various policies were designed to ensure fast execution of the projects and the reduction – or removal – of conditions that could potentially delay the project. Environmental clearance in the new millennium was identified as one of the roadblocks in the pathway of fast execution of projects, which thus required *re-engineering* (GoI/MoF 2009; Nandimath 2009). Within this world-view, environmental clearance and EIA follow-up are seen as another licence to be procured before the project can take off. It is simply a hurdle to be crossed (Rajaram and Das 2006) and the project will at the maximum stick to the minimum. If possible, it will even bypass legal requirements.

Since EIA and EIA follow-up are largely regulator driven, the regulatory agencies, whether the MoEF or the Environment and Forest Departments in various states, have to play a proactive role. But the various state governments are competing against each other for attracting more and more private capital. Thus, they have less and less incentive to perform the role of a strict environmental regulatory agency. For many states, hydropower has been identified as an 'avenue for revenue generation' (quote by a senior power department official interviewed in May 2009).

Thus, any strong regulatory action would be in sharp contrast to the overall policy in favour of hydropower-based revenue generation. Under this macro-political-economic scenario, it is not surprising that the Himachal Pradesh High Court heavily criticises the Himachal Pradesh Government for failing to perform its duty as a custodian of natural resources.

The review of the Annual Reports of the MoEF over the last 25 years shows that EIA follow-up has, at best, been reduced to compliance monitoring, with little or no effects monitoring taking place. But, as the Shukla Committee report and the orders of the High Court of Himachal Pradesh show, often environmental baselines are under-reported, either deliberately or due to a genuine lack of environmental information. This implies that mitigation measures may not automatically translate into the mitigation of adverse impacts. Under these circumstances it is minimisation of adverse impacts – and hence effects monitoring – which is more important than merely conducting desk reviews of the six-monthly compliance reports.

Generation of knowledge, and thus learning, or feedback, is one of the greatest advantages of proper design and implementation of EIA follow-up. But the learning takes place only when effects are monitored and evaluated, and most importantly, when remedial actions are then taken up (adaptive management). The review of Annual Reports of the MoEF hardly provides any information on the *learning* front. Rather, the Shukla Committee constituted by the High Court of Himachal Pradesh provides insights on how a field-based review can provide better short-term managerial decisions (*short-term learning*).

While EIA provides an environmental licence, EIA follow-up could provide a social licence to continue with the activity through enhancement of mutual trust and through the incorporation of different public involvement mechanisms in the follow-up process (Arts et al. 2001). In the case of India, apart from EIA Notification 2006 provisions for putting up environmental clearance and six-monthly compliance reports in the public domain, there is little public involvement in the form of communication or participation. The Annual Reports do not provide information on the public involvement in the EIA follow-up process. The case study of the Allain Duhangan Project shows that gaps in EIA follow-up can actually hurt a project by reducing the mutual trust between the project proponent and the public and may thus reduce the legitimacy of the project.

Experience from EIA follow-up in other parts of the world showed that the presence of a regulation tends to strengthen the EIA follow-up. In the case of India, it was seen that although in the early 1990s – in absence of any legal framework – the MoEF had defined a broad scope for EIA follow-up, over the years the scope was narrowed down. Various clauses which would have strengthened the EIA follow-up, and which were a part of EIA Notification 1994, got diluted under the 2006 EIA Notification. These weaknesses imply that the regulatory framework, which is usually identified in the EIA follow-up literature as an important contextual variable (Morrison-Saunders et al. 2003), has also contributed to the gradual decline of EIA follow-up in the last three decades. Thus, on one hand, the coverage of EIA has increased, with more and more projects receiving environmental clearance. On the other hand, the institutional strengthening that was required within EAC, MoEF and its regional offices to strengthen the EIA follow-up has been missing. This is the *institutional reason* behind the consistent decline of environmental governance in general and EIA follow-up in particular.

In recent times, there have been some efforts promoted by the MoEF to revamp the entire environmental governance system, and there have been some commendable public involvement efforts in that regard. But based on the analysis of this chapter, it can be argued that the intended reform of bringing in new institutions in place of the old ones (GoI/MoEF 2009b, 2010b) can only address the *institutional constraints* component of the general decline. To strengthen environmental governance and EIA follow-up in the country more generally, the people's image of what *environment* is, and what its significance is, has to change. This is a far bigger challenge.

Acknowledgements The author acknowledges the generous support of International Postgraduate Studies in Water Technologies (IPSWaT) for supporting the field research for this study. The author would also like to thank the many resource persons, interactions with whom have enormously helped the author in developing a better understanding of environmental governance in India. The author would like to thank an anonymous reviewer whose comments have significantly improved the chapter.

References

Arts J, Meijer J (2006) Designing for EIA follow-up: experiences from Netherlands. In: Morrison-Saunders A, Arts J (eds) Assessing impact handbook of EIA and SEA follow-up. Earthscan, London, pp 63–96

Arts J, Morrison-Saunders A (2006) Theoretical perspectives on EIA and EIA follow-up. In: Morrison-Saunders A, Arts J (eds) Assessing impact handbook of EIA and SEA follow-up. Earthscan, London, pp 22–41

Arts J, Nooteboom S (1999) Environmental impact assessment monitoring and auditing. In: Petts J (ed) Handbook of environmental impact assessment. Vol 1: environmental impact assessment: process, methods and potential. Blackwell Science, Cornwall, pp 229–251

Arts J, Caldwell P, Morrison-Saunders A (2001) Environmental impact assessment follow-up: good practice and future directions – findings from a workshop at the IAIA 2000 conference. Impact Assess Proj Apprais 19(3):175–185

Cashmore M (2004) The role of science in Environmental Impact Assessment: process and procedure versus purpose in the development of theory. Environ Impact Assess Rev 24: 403–426

Chauhan K (2009) Forest Dept's [Department's] role under scanner. The Tribune. http://www.tribuneindia.com/2009/20091124/himachal.htm#10. Accessed 12 July 2010

Choudhury N (2010) Sustainable dam development in India – between global norms and local practices (Discussion paper 10/2010). German Development Institute, Bonn

GoI/MoEF (Government of India, Ministry of Environment and Forest) (1987) Annual report 1986–87. Ministry of Environment and Forest, Delhi

GoI/MoEF (Government of India, Ministry of Environment and Forest) (1988) Annual report 1987–88. Ministry of Environment and Forest, Delhi

GoI/MoEF (Government of India, Ministry of Environment and Forest) (1989) Annual report 1988–1989. Ministry of Environment and Forest, Delhi

GoI/MoEF (Government of India, Ministry of Environment and Forest) (1990) Annual report 1989–1990. Ministry of Environment and Forest, Delhi
GoI/MoEF (Government of India, Ministry of Environment and Forest) (1991) Annual report 1990–1991. Ministry of Environment and Forest, Delhi
GoI/MoEF (Government of India, Ministry of Environment and Forest) (1992) Annual report 1991–1992. Ministry of Environment and Forest, Delhi
GoI/MoEF (Government of India, Ministry of Environment and Forest) (1993) Annual report 1992–1993. Ministry of Environment and Forest, Delhi
GoI/MoEF (Government of India, Ministry of Environment and Forest) (1994a) Annual report 1993–1994. Ministry of Environment and Forest, Delhi
GoI/MoEF (Government of India, Ministry of Environment and Forest) (1994b) Environment Impact Assessment notification, S.O. 60(E) (27 January 1994). In: Nandimath OV (ed) Oxford handbook of environmental decision making in India: an EIA model. Oxford University Press, New Delhi, pp 174–183
GoI/MoEF (Government of India, Ministry of Environment and Forest) (1995) Annual report 1994–1995. Ministry of Environment and Forest, Delhi
GoI/MoEF (Government of India, Ministry of Environment and Forest) (1996) Annual report 1995–1996. Ministry of Environment and Forest, Delhi
GoI/MoEF (Government of India, Ministry of Environment and Forest) (1997) Annual report 1996–1997. Ministry of Environment and Forest, Delhi
GoI/MoEF (Government of India, Ministry of Environment and Forest) (1998) Annual report 1997–1998. Ministry of Environment and Forest, Delhi
GoI/MoEF (Government of India, Ministry of Environment and Forest) (1999) Annual report 1998–1999. Ministry of Environment and Forest, Delhi
GoI/MoEF (Government of India, Ministry of Environment and Forest) (2000) Annual report 1999–2000. Ministry of Environment and Forest, Delhi
GoI/MoEF (Government of India, Ministry of Environment and Forest) (2001) Annual report 2000–2001. Ministry of Environment and Forest, Delhi
GoI/MoEF (Government of India, Ministry of Environment and Forest) (2002) Annual report 2001–2002. Ministry of Environment and Forest, Delhi
GoI/MoEF (Government of India, Ministry of Environment and Forest) (2003) Annual report 2002–2003. Ministry of Environment and Forest, Delhi
GoI/MoEF (Government of India, Ministry of Environment and Forest) (2004) Annual report 2003–2004. Ministry of Environment and Forest, Delhi
GoI/MoEF (Government of India, Ministry of Environment and Forest) (2005) Annual report 2004–2005. Ministry of Environment and Forest, Delhi
GoI/MoEF (Government of India, Ministry of Environment and Forest) (2006a) Annual report 2005–2006. Ministry of Environment and Forest, Delhi
GoI/MoEF (Government of India, Ministry of Environment and Forest) (2006b) The 2006 notification, S.O. 1533(E) (14 September 2006). http://www.moef.nic.in/legis/eia/so1533.pdf. Accessed 7 Apr 2011
GoI/MoEF (Government of India, Ministry of Environment and Forest) (2007) Annual report 2006–2007. Ministry of Environment and Forest, Delhi
GoI/MoEF (Government of India, Ministry of Environment and Forest) (2008) Annual report 2007–2008. Ministry of Environment and Forest, Delhi
GoI/MoEF (Government of India, Ministry of Environment and Forest) (2009a) Annual report 2008–2009. Ministry of Environment and Forest, Delhi
GoI/MoEF (Government of India, Ministry of Environment and Forest) (2009b) Towards effective environmental governance: proposal for a National Environment Protection Authority (Discussion Paper). Ministry of Environment and Forest, Delhi
GoI/MoEF (Government of India, Ministry of Environment and Forest) (2010a) Annual report 2009–2010. Ministry of Environment and Forest, Delhi

GoI/MoEF (Government of India, Ministry of Environment and Forest) (2010b) Reforms in environmental governance: with special reference to establishment of National Environment Assessment and Monitoring Authority (NEAMA) (Discussion Paper). Ministry of Environment and Forest, Delhi

GoI/MoEF (Government of India, Ministry of Environment and Forest) (2011) Annual report 2010–2011. Ministry of Environment and Forest, Delhi

GoI/MoF (Government of India, Ministry of Finance) (2009) Expert group to examine the schemes of statutory clearances for industrial and infrastructure projects in India. http://www.indiaenvironmentportal.org.in/files/Report%2020Ministry%20of%20Finance.pdf. Accessed 2 Sept 2009

HCHP (High Court of Himachal Pradesh) (2009) Civil Writ Petition Public Interest Litigation No. 24/2009: Court on its own-versus-state of Himachal Pradesh. Interim Order (23 December 2009)

HCHP (High Court of Himachal Pradesh) (2010a) Civil Writ Petition Public Interest Litigation No. 24/2009: Court on its own-versus-state of Himachal Pradesh. Interim Order (09 August 2010)

HCHP (High Court of Himachal Pradesh) (2010b) Civil Writ Petition Public Interest Litigation No. 29/2009: Court on its own-versus-state of Himachal Pradesh. Interim Order (06 August 2010)

HCHP (High Court of Himachal Pradesh) (2010c) Civil Writ Petition Public Interest Litigation No. 29/2009: Court on its own-versus-state of Himachal Pradesh. Interim Order (09 August 2010)

IFC/CAO (International Finance Corporation, Compliance Advisor Ombudsman) (2004) Assessment report complaint regarding Allain Duhangan hydropower project Himachal Pradesh, India. Office of the Compliance Advisor Ombudsman International Finance Corporation Multilateral Investment Guarantee Agency, Washington, DC

IR (International Rivers) (2009) International Rivers. Website: http://www.internationalriversorg/n/outh-asia/ndia/affected-people-world-bank-regarding-allain-duhangan-hydro-project. Accessed 2 Sept 2000

Jha-Thakur U, Fischer TB, Rajvanshi A (2009) Reviewing design stage of Environmental Impact Assessment follow-up: looking at the open cast coal mines in India. Impact Assess Proj Apprais 27(1):33–44

Marshall R (2005) Environmental Impact Assessment follow-up and its benefits for industry. Impact Assess Proj Apprais 23(3):191–196

Marshall R, Arts J, Morrison-Saunders A (2005) International principles for best practice EIA follow-up. Impact Assess Proj Apprais 23(3):175–181

Menon M, Kohli K (2008) Reengineering the legal and policy regimes on environment. Econ Political Wkly 43(23):14–17

Menon M, Kohli K (2009) From impact assessment to clearance manufacture. Econ Political Wkly 44(28).20–23

Morrison-Saunders A, Arts J (2005) Editorial. Learning from experience: emerging trends in environmental impact assessment follow-up. Impact Assess Proj Apprais 23(3):170–174

Morrison-Saunders A, Arts J, Baker J, Caldwell P (2001) Roles and stakes in Environmental Impact Assessment follow-up. Impact Assess Proj Apprais 19(4):289–296

Morrison-Saunders A, Baker J, Arts J (2003) Lessons from practice: towards successful follow-up. Impact Assess Proj Apprais 21(1):43–56

Nandimath OV (ed) (2009) Oxford handbook of environmental decision making in India: an EIA model. Oxford University Press, New Delhi

Paliwal R (2006) EIA practice in India and its evaluation using SWOT analysis. Environ Impact Assess Rev 26:492–510

Patton MQ (2002) Qualitative research & evaluation methods, 3rd edn. Sage, Thousand Oaks

Rajaram T, Das A (2006) EIA in India. Need for participatory and sustainable principles in India's EIA system: lessons from the Sethusamudram Ship Channel Project. Impact Assess Proj Apprais 24(2):115–126
Ramesh J (2010) The two cultures revisited: some reflections on the environment-development debate in India. Jawaharlal Nehru Centre for Advanced Scientific Research, Bangalore
Rowe G, Frewer L (2005) A typology of public engagement mechanisms. Sci Technol Hum Values 30(2):251–290
Roy D (2008) Hydropower in Uttarakhand: is 'Development' the real objective? Econ Political Wkly 43(41):19–22

Chapter 7
Rivers, Dams and Landscapes: Engaging with the Modern on Contested Grounds

Hanna Werner

> *Nowhere have the conflicts between [...] competing visions of development been more vividly displayed than in the intensified transnational contestation over big dam projects*
>
> (Khagram 2004, p. 4).

Abstract This chapter discusses one of the most disputed topics in Indian environmental history, namely, the construction of large-scale hydroelectric dams in the context of post-independent developmental politics. It deals with large dams as part of twentieth-century planning in India and the 'ideological designs', on which their implementation was based against alternative visions and an emerging opposition. It seems reasonable to assume that infrastructural artefacts such as large dams, apart from their role and function on a national agenda, are an attempt to allot resources and counter the unpredictability of nature through engineering efforts. With regard to social resistance and alternative development models, existing links between this implementation and modernity theory are discussed for the Indian context.

Keywords Large dams • Development discourse • Modernity • Social visions • Resistance • Tehri Dam • India

Symbols of modernity or signs of social disparity, indicators of economical progress or issues of ecological concern – large dams have been at the centre of developmental discourses in India at least since the 1980s. Regardless of their

---

H. Werner, M.A. (✉)
Institut für Asien- und Afrikawissenschaften, Seminar für Südasien-Studien, Humboldt-Universität zu Berlin, Unter den Linden 6, D-10099 Berlin, Germany
e-mail: wernerha@hu-berlin.de

M. Nüsser (ed.), *Large Dams in Asia*, Advances in Asian Human-Environmental Research, DOI 10.1007/978-94-007-2798-4_7,
© Springer Science+Business Media Dordrecht 2014

political evaluation, large dams are obviously not 'neutral' technologies. Their implementation is closely intertwined with larger visions of appropriating nature and optimising its resources for human need and want.[1] As gigantic manifestations of state power, dams can be seen as visualising no less than a nation's dream of centralising and controlling its resources – and demonstrating it to the rest of the world.[2]

Focusing on India, this chapter deals with large dams as part of post-independent development planning and the 'ideological designs' on which their implementation was based, in the light of alternative societal visions and emerging resistance. It is suggested that one can understand large and mega dams as icons of a technocratic modernisation imperative, already envisaged in the colonial project of British rule and further employed within the (ideological) framework of Indian nation-building.[3] A lot has been said about the symbolic quality of large dams as icons of modernity, hence I will not go into more detail here.[4] Yet a crucial aspect with regard to the transnational historical setting seems worthwhile mentioning: The era of building hydroelectric mega dams began in the USA in the first half of the twentieth century. While the current trend in the USA and in large parts of Europe is to decommission a number of projects due to obvious negative ecological consequences, great parts of the so-called developing world still place their trust in large-scale technologies to promote their country's economic progress.[5] This trend, however, has not remained unchallenged. With the number and size of project implementations and an increased public awareness of the adverse consequences, scepticism and resistance have grown. Pro- and contra-arguments regarding large dams are well known[6]:

---

[1]See the reflections on historical shifts in technology imaginations in Sect. 7.2 of this chapter.

[2]The fact that a lot of projects are the result of private-public partnerships and financed by external agencies does not necessarily diminish the display of national power and prestige that is achieved through their construction. Yet it impacts the state's scope of action to a considerable extent.

[3]See e.g. D'Souza (2006) and Klingensmith (2007). Klingensmith excellently illustrates how the accomplishment of a specific development paradigm was inseparably connected to the success of the nation. As he correctly states, it is important to recognise the historicity of the idea of the nation that, though not old, has nonetheless succeeded in becoming one of the most incommensurable 'designs' of modern history (see Klingensmith 2007, pp. 20f). For a convincing study of the political implications of irrigation and large dams in British and post-independent India, see also Singh (1997).

[4]See Chap. 1 of this volume.

[5]Compare the table in the World Commission on Dams report (WCD 2000, p. 8). The statement could be admittedly read as a naïve fallacy, since the mere fact that new projects are no longer pushed forward in Europe and North America is not necessarily a reasonable assessment of its developmental politics, nor does it conceal the fact that a lot of so-called developmental aid backs and patronises 'outdated' large-scale hydrological infrastructure.

[6]The most popular accounts include McCully (1996), Goldsmith and Hildyard (1984), but there are of course various sources, monographs as well as articles in scientific and other journals. For India, *Economic and Political Weekly*, *Down to Earth*, *Outlook*, *Frontline* and *Himal* are just few of the sources that have repeatedly reported on large dams and accompanied the various struggles on regional as well as global levels.

Besides ecological consequences, social issues, namely, involuntary displacement, are amongst the major arguments raised by the opponents' side. Acknowledging the numerous cost-benefit analyses that illustrate the advantages and disadvantages of large dam building, one must keep in mind the limitations of such an approach. Damming a river has implications that transcend the 'materiality' of the socio-economic and ecological sphere. The way in which nature is appropriated and resources are (re-)distributed reflects the political constitution and social vision of the respective country. The subject therefore requires a critical reflection on the conceptional definitions of development and the role of knowledge hierarchies in decision- and policy-making processes. Such an approach queries the very core of nation-building and the foundational principles of societal organisation itself. To quote Arundhati Roy: 'What is at issue now is the very nature of our democracy. Who owns this land? Who owns its rivers? Its forests? Its fish?' (Roy 1999, n. p.)[7]

This chapter seeks to provide – preliminary – answers to two ostensibly simple questions that one nonetheless fails to address, if the analysis remains restricted to cost-benefit considerations on the performance of large-scale hydroelectric projects. First, if it is obvious that large dams are more often than not ecologically unsound, socially unjust and economically inefficient, why would a nation like India continue to invest enormous sums in their construction? Taking into consideration that those projects are often upshots of public-private partnerships with a clear allotment of shares and profits, one could easily subscribe to a straightforward utilitarian answer. A deeper historiographical analysis and deconstruction of the transnational setting, however, shows that this type of argument falls short of addressing the question of ideology, which plays an important role in any developmental discourse and its infrastructural manifestations.

The second, somewhat related question addresses the issue of ensuing resistance against large projects: Drawing on the observation of recent struggles, one could ask why decades of protest, though very effective in certain situations, have not essentially succeeded in questioning the core parameters of progress-oriented planning in the context of national resource management, namely, its persisting efficacy as instrument and driving force of sociability.[8] Rather than presenting a critique of 'developmentalism' (Dirlik 2002, p. 20), this chapter attempts to take hold of the trajectory of its historical realisation as a predominant principle of social organisation, which seems to have resulted in an incontestability of its legacy. I suggest that this is one of the major reasons why the contested discourse over

[7]This question of redistribution is anchored in a hegemonic discourse that defines how resources are to be used and implies a development paradigm that tends to privilege sections of society that are anyhow part of the national 'modernity project'.

[8]Being critical of common developmental paradigms is of course not a structural prerequisite for the articulation of opposition against large dams. The impact and political subtext of such a criticism, however, is entirely different. See the last section of this chapter.

large dams has often been limited to performance-oriented cost-benefit reflections. Taking for granted the key arguments of a growth-oriented development pattern, the majority of critical accounts would suggest, that '[...] major projects will continue to be built because of the increasing worldwide demand for food, potable water, energy, and economic development. The real challenge is to implement workable measures to avoid the environmental, social, and institutional pitfalls of such projects' (Bizer 1987, p. 757). What is easily ignored is the fact that such an argument does not convey a technical, but an utterly political choice of how to go about social change. As convincing as it might sound in the light of an evocation of the 'pressing needs of the present', a reformist approach with regard to socioecological change, which does not touch upon the self-narrated evidence of developmental paradigms, is just one of historically contingent interventionist options.

This chapter is divided into five parts. Starting with a contextualisation of terminologies, it deals with the transformation of nature-technology relations that accompanied the consolidation of modernity and its monumental manifestations in the early twentieth century and takes a look at the 'southwards spread' of its paradigms. Against this background, light is shed on the historical context of dam building in post-independent India. 'Technocratic' development approaches that determine water management patterns are confronted with alternative notions of nature and landscape, in particular, water bodies. As an exemplifying case study, the North Indian Tehri Dam Project illustrates the complex intertwinement of nature/landscape perceptions and national resource management in India. The chapter concludes with some remarks on the historical and contemporary shape of the large dam debate and explores spaces for (discursive) intervention.

7.1 Contextualising Terminologies

> Large dams are one of the many features of the modern world. We can gather a lot of evidence against dams, but then we can do the same for other features of the modern world that pass for 'development' – nuclear power plants, megalopolises, exploding automobile population, conversion of wetlands into urban centers, and so on.
>
> If we say 'No' to big dams, we should be prepared to say 'No' to those other things as well. But this requires a rethinking of what we see as 'development' [...]. (Iyer 2006, n. p.)

The question of whether a dam will be built or a new project should be implemented is not merely a rational verdict for 'the greater common good' (Roy 1999), but rather a decision on the implementation of the very idea of development itself. Habitually, the paradigm associated with growth and progress is considered as modern.[9] It is important to keep in mind that modernity itself reflects an epochal as well as a theoretical concept that has varied over time. Thus, a critique of modernity does not only require thorough historiographical (re-)reading, it also calls for the deconstruction of its criteria or 'metanarratives'.[10] 'Modern experience' at the turn of the twentieth century reflects a major change in the perception of time and space that manifests itself as a shift in spatial perspective, increased mobility, experience of contingency and tendentially unlimited possibilities. Since modernity draws its legacy from a universalist claim that includes the intrinsic production of normativity and a perception of the future that is based on the experienced present, it envisages a radically innovative point of departure for its projects (see Klotz 1999, p. 17). Those characteristics have formed the precondition for a constructivist or rather technocratic appropriation of the world. The emerging concept of progress is probably its most explicit product. Infrastructural monuments are a visual landmark of its paradigmatic success. In that sense dams are not only political symbols of modernity. As architectural artefacts they are also manifestations of the aesthetics of modernity.[11] Against this background, opting for large-scale projects appears in a particular light: 'Developmental monumentalism is a way for a country to announce to the world: We have arrived, we are modern' (Baruah 2006, p. 62).

As a multilateral concept to balance economic asymmetries – within the nation and beyond its borders – the term development was for the first time strategically employed in the 1950s in the USA. Emerging in a post-war production set-up, this conception was closely connected with the notion of economic growth. The bureaucratic language of techno-economical achievement, however, conceals the historical heritage of the concept 'development' that was and still is far from being politically neutral. The implementation of the metanarratives of European modernity within the overarching vision of a worldwide 'catch-up' syndrome gave rise to the adaptation of a top-down concept of development. Shifting the focus to the history of large dams in India displays an analogous picture. Post-independent

[9]Development, here, is understood as one of the central paradigms that accompany the striving for modernity since the midst of the last century.

[10]This suggestion is reflected in the definition of postmodernity: a later, reflexive and revisionist stage of modernity itself that questions particularly those metanarratives on which its creation and inherent legitimisation had uncritically been built upon. A crucial example is the scepticism that historical contingency is necessarily problematic and must therefore be overcome by putting into place technocratically motivated structures (see Klotz 1999, p. 24 and Makropoulos 2004, p. 369).

[11]Since the 'era of large dam building' started in the first half of the twentieth century in the USA, particularly during the time of the New Deal with the *Tennessee Valley Authority* (TVA) and other projects, a large body of literature on dam building in the USA is available. A lot of these works deal less with cost-benefit considerations, but rather focus on political and aesthetical implications that accompanied early ideas of development implementation and the ideology of building big projects. See e.g. Wilson (1985) or Worster (1985).

development has largely been considered as a top-to-bottom process, burdened with the colonial inheritance of the ideology of techno-political omnipotence: '[...] since 1947, governments in India, building on a destructive colonial legacy and twentieth-century modernist ideology, have aggressively sought to ascertain and meet water demands through either big-engineering projects or intensive extraction technologies rather than concentrating on localised conservation efforts or on strengthening indigenous water knowledge traditions' (D'Souza 2008, p. 120).

It would be nevertheless misleading to depict the adaptation of 'Western' ideas as a one-dimensional process and to assume that developing states, or rather the agents of their political agenda, have been merely at the receiving end of a one-way paradigm transfer.[12] The claim for the historical origins of the project modernity can be found in Europe. That, however, neither explains the conventional usage of this origin as a timeless global frame of reference nor justifies its 'civilising mission'.[13] With the critique of linear modernisation theory, a revised concept of modernity has come to pass that transcends the idea of being solely a Western product. It has been acknowledged that a lot of knowledge, which shaped modernity's face in Europe, was based on experiences and experiments made in its 'overseas laboratories' – resulting in transnational flows of meaning that have ever since required processes of reinterpretation and improvisation. Likewise, 'Western thought' has not just been uncritically adopted by post-colonial states, but modified and revised according to the respective situational requirements. This observation has a significant impact on the discussion of the logic and ideology of the development paradigm manifested in large dams as 'monuments of modernity' and requires a further exploration of the historical circumstances. Historical screenplays are allegedly contingent – yet they are not coincidental. Dealing with the historical possibilities of the implementation of alternative visions for a post-independent India, one has to subsequently consider various interdependent factors, stretching from biographical trajectories and commitments to transnational engagements.[14]

[12]That would reflect an assumption Klingensmith refers to as 'polemical criticism' (2007, p. 30). It would be equally naïve to assume that every aspect of contemporary Western societies is automatically modern with reference to European modernity's own standards (see Chatterjee 1997, p. 264).

[13]For a convincing critique of this ethnocentric approach towards modernisation theory and historiography, see Chakrabarty (2000). Kaviraj persuasively lists a number of 'adaption modes' – the 'translation' of practices, specificities of sequence, improvisation and reflexivity – that make the assumption of a linear, ethnocentric modernisation theory highly unlikely. In analogy to Chakrabarty's use of the adjective 'provincial', he employs the term 'parochial' (2005, pp. 516ff).

[14]An important aspect is the exposure to foreign education systems, 'Western' nationalist thought, and so forth.

7.2 Historicising Artefacts

> As I walked round the site (Bhakra-Nangal) I thought that these days the biggest temple and mosque and gurudwara is the place where man works for the good of mankind. [...] When we see big works, our stature grows with them, and our minds open out a little (Nehru 1954/1958, p. 3).

> For some time past, however, I have been beginning to think that we are suffering from what we may call, 'disease of gigantism'. We want to show we can build big dams and do big things. This is a dangerous outlook developing in India [...] (Nehru (1958), quoted from Thakkar 2005a, p. 4).

> Bhakra, the new temple of resurgent India, is the symbol of India's progress (Nehru (1963), quoted from ibid., p. 1).[15]

Three statements by the then Prime Minister Jawaharlal Nehru that could not be more disparate.[16] Remarkably, solely the statements depicting dams as 'temples of modernity' have gained attention in the mainstream discourse. Those are also the ones that have made it into the canon of education (see Roy 2000). It seems that ever since Nehru's statements have settled in the public discourse, an affirmative reception of large dams has become highly conclusive.[17] To date, Nehru is epitomised as the science adherent technocrat, yet driven by a socialist strive for the greater common good. The idea of scientific achievements on Indian ground was not only motivated by the competitive endeavour to catch up with Western modernisation; it was rather an emancipatory vision that did not simply employ external parameters, but aimed at syncretising colonial heritage – its infrastructure, not its legacy – with Indian resources. Efficient capacity building through the usage of rural resources though was seen as an improper method to achieve twentieth-century economic goals, including the eradication of large-scale scarcity in the countryside. The prophets of a technology-based, scientifically legitimised progress of the Indian state planted their trust in centralised resource management, and industrialisation was seen as the foremost means to alleviate the Indian people's state of starvation and 'backwardness'.[18]

[15]The quotations are taken from the following speeches Jawaharlal Nehru's: 1954 at the Nangal canal, 1958 at the 29th annual meeting of the *Central Board of Irrigation and Power* and 1963 at the inauguration of Bhakra Project.

[16]It has been critically assumed that Nehru changed his mind over time. Yet looking at his emphatic support of dams as an element of large-scale industrialisation throughout his life time, there is not much evidence that would suggest such a change. For a comment on 'Prime Ministers and big dams', see Guha's article in *The Hindu*, 18 December 2005.

[17]This has significant implications for the articulation of any critical standpoint; see the concluding section of this chapter.

[18]Though Nehru's vision is particularly emphasised here, he was by far not the only advocate of large dams and heavy industrialisation as means to promote the country's economic development. To understand the prevalent mindset of late pre- and post-independence 'nation-building circles', one has to contextualise the biographies of personalities like M. N. Saha, S. Sen, A. N. Khosla, K. L. Rao and others, who have been intellectually and practically involved in setting up large dam building in India. For a detailed account of their role, see Klingensmith (2007).

Gandhian thought, 'which was completely opposed to Western industry and technology-driven modernity, was considered to be backward-looking' and seen as a hindrance to rapid social change and growth (Chatterjee 2007, p. 46). 'Father of the Nation' for many, Gandhi has ever since frequently been characterised as 'antimodern', 'antiscience'[19] and therefore, notably, even as 'anti-national'. Yet rejecting Gandhi's ideas on the basis of their characterisation as 'traditional' does not only imply a static notion of his philosophy, it also sets clear demarcations to his position in the nation-building project. Because of his outstanding role in the Indian independence struggle and the immense respect he gained within the Indian population, the integration of Gandhi's heritage into post-colonial India was nevertheless inevitable. The result appears as an almost ceremonial integration of Gandhi's moral philosophy into the political rhetoric of the (early) nation, while the undesirable components of his thinking, like the strict opposition to the centralised state or his objection to modern machinery, were either disposed of or adjusted to the project of nation-building (see Chatterjee 2007, p. 46). Conversely, a major part of the socioecological movements in India in the twentieth century – anti-dam agitation just being a comparatively late outcome of it – draw their origins from Gandhian ideas and performances.

Evoking Nehru and Gandhi as two epigones incorporating opposing visions for a post-independent India suggests an unreasonably simplified picture of the historical configuration. Yet since 'Nehru represented, unlike Gandhi or any other leader, not only a clean break with tradition but the symbol of a new era, where poverty would soon be left behind in the face of the relentless onslaught of industrialization' (D'Monte 1985, p. 1), it seems reasonable to 'extract' some of the arguments his approach stands for to disclose and discuss the (contingent) historical political choices that have been made. I consider the post-independent technocratic view on planning and accessibility of natural resources as particularly important for a better understanding of the landmarks of the historical set-up. As an underlying ideology, it has set parameters that are still viable today and have shaped the discourse and its 'material' effect, namely, the justification of large-scale projects and interventions. The justification of the use of the term 'technocracy' with regard to water resource

[19]A number of authors suppose that this view is incorrect (see e.g. Sahasrabudhey 2002; Prasad 2001). Prasad e.g. emphasises that Nehru himself, in stressing Gandhi's role as a religious person, has added to the public opinion and strengthened the divide between the two approaches. The author further states that while Nehru's views on science and industrialisation have been studied intensely, there has been a severe shortage of studies on Gandhi in the sphere of science policy studies as well as lack of dealing with the issue of science in Gandhian studies. This has added to the notion of Gandhi as 'antiscience' and created an exclusive opposition that narrowed down the space for explorations on 'alternative' science models suggested by Gandhi. Authors such as Ashis Nandy have repeatedly stated that it was rather the ideology of science, respectively, technology, manifested in a civilising mission without 'morality' that Gandhi criticised, than its implementation per se (see Prasad 2001, pp. 3721f).

management is concisely pointed out by a statement Meghnad Saha, physicist and influential mentor of the *Damodar Valley Corporation* (DVC), made in front of the Parliament in 1955: 'At one time floods and famines were considered as acts of God, meant to punish the people for his sins. That stage is over now. It has been found that the floods and famines can be efficiently tackled by taking proper scientific measures' (Saha 1955, p. 17).[20]

One recognises that not only the role of science, which is expected to have an important function in accomplishing societal goals, but also the element of (social) control is strongly emphasised. While the example given illustrates aspired control over nature, the implications of such 'security measures' are obvious. Technocracy as a socio-philosophical concept implies the imaginary as well as the material conquest of nature. The objectification of nature leads to the essentialist differentiation between culture and nature that characterises the (self-)description of modern societies. Rather functionalist, nature is seen as a container of resources for humankind, a form that can be made operable for human needs. The idea of limitless mastery has yet another side to it: The experience of a horizon of limitless possibilities is burdened with a large amount of insecurity in so far as the predictability of history (making) – and its monumental outputs – has vanished. The technocratic obsession for control is therefore not only a functional but also a cognitive condition, which is based on the threatening experience of contingency that has accompanied human creation ever since a constructivist worldview has been adopted. Following this strand of ideas, one is tempted to assume that monumental artefacts such as large dams, apart from their role and function as national icons, are an attempt of humankind to counter the unpredictability of nature. This is where the dilemma, besides dams having a number of immediate adverse consequences, might eventually be located. A recent review of van Laaks' 'White Elephants' evokes this idea quite eloquently: Megaprojects 'are wrong not merely because huge projects typically fail, leading to environmental and human catastrophe or falling far short of their initial promise, but also because they reflect the twentieth century's underlying hubris: the delusion of an unlimited capacity to remake nature and society and a

[20]This affirmative view of technology and science, however, does, at least in the case of Saha, by no means imply a blind faith in 'Western' expertise. Saha continues his speech: 'I think the Ministry should do well to pay a little attention to these things [his studies about hydrological condition of the river, necessity of dams on the tributaries, H. W.] From time to time they are calling great foreign experts to find out whether these projects are working well or not working at all. I do not understand the sense in calling these experts. The foreign experts, however great they may be in their own country, are not acquainted with the physical conditions in this country. Most of them have no idea of the Monsoons; most of them have no idea of the precipitation in the different river valleys. These points require an intensive study by our own people, and when foreign experts are invited from other countries, I do not see why they should not ask those people who have made a little study of these things for their opinion' (1955, pp. 18f). Yet when it comes to watershed development, a lack of qualified personnel and sufficient hydrological surveys including the required 'machinery' is acknowledged, and Saha warns the government to avoid hasty decisions and calls attention to the long journey ahead (ibid.).

naïve and unwarranted faith in salvation through technology – in short, the attempt to be God' (Gispen 2000, pp. 1001f).[21]

The idea of infinite malleability of nature is relatively new; its origins do not reach far beyond the turn of the twentieth century. This suggests that there is a fundamental difference in the functionality of dams throughout history, which indeed implies a change of paradigm. While the earliest beginnings of dam building date back to 3000 B.C., the project of hydroelectric large dam building is a phenomenon of the modern times, expanding with the peak of industrialisation in the nineteenth century. From the middle of the twentieth century onwards, one can speak of an actual era of large dams; it emerged in the USA, shortly afterwards took off in the then Soviet Union, before it expanded southwards (see McCully 1996, pp. 13ff).

By turning our attention towards India we learn that its history of dam building is not solely a post-independence project, although most of India's large dams have been built in the second half of the twentieth century. During British colonial rule, a couple of hundred dams were built, mainly for irrigation of farmland in the light of an increasingly export-oriented agriculture. The fact that pre-independence visions have quite often formed the basis of later projects is especially noteworthy in this context. Those 'irrigation visions' differed in many ways from pre-existing structures. It was remarkably not only the technology (of the then canal infrastructure) that was subject to modifications; of greater significance are the administrative and political changes behind the scene. The new structures implemented by the British were in many ways 'delocalised' and facilitated their centralised bureaucratic control, since they did not require community participation at the high level that was essential for 'traditional' structures (Sengupta 1993, p. 10). I will come back to this aspect in the concluding section of this chapter, with a deliberation on a potential revival of 'traditional' infrastructure.

The non-participatory water management approach induced by the colonial administration seems to have been largely compatible with the interests of the national government. The reception of dams in India at the time of early post-independence was quite passionate: 'If it did not create the initial enthusiasm for dams, the Government of India certainly responded to and encouraged it, by highlighting them as the key to India's future' (Klingensmith 2007, p. 260). It is especially after independence that large dams were depicted as symbols of progress and modernity, later accompanied by applied optimisation programmes in the agricultural sector during the *Green Revolution*.[22]

As mentioned above, it is not merely the technology of dam building, which resulted in an improvement of infrastructure and efficiency that has changed over time, rather the perception of technology and its function itself have been transformed. Large dams might have been built more than 3,000 years ago – as

---

[21]That is what James Scott describes as the ideology of 'high modernism' (1998, pp. 4f).

[22]The main goals were improvement of yields and food export, on the one hand, prevention of droughts on the other.

large-scale irrigation structures, as prestige objects. Yet I presume that their implications with regard to the underlying perception of nature as an operable utensil of human planning and control have changed significantly. A closer look at the historical transformation of the perception of technology illustrates this argument: The premodern understanding of technology drew upon the ancient idea of 'techné', which suggests that natural perfection can be reached by its imitation. Technology was seen as an enhancement of natural conditions and a partial substitute of its deficiencies. In modern times, technology paired with science gains a significantly different quality. Since it reduces nature to the substance of constructivist endeavours, it implies the appropriation of its resources, its conquest. Technological aspirations have transgressed the boundaries of experience and thus the limits nature had previously set. Hence, the role of technology no longer consists in the imitation of nature, but in its overcoming, its optimization through its replacement (see Makropoulos 2000, pp. 8f).[23] The assertion of a culture-nature divide as one of modernity's essential characteristics has had its impacts and led to a condition, in which nature has seldom been seen as more than a malleable agglomerate of resources, available and accessible for human intervention. Criticising this approach as a modern condition, distinctive for Western societies and their voracious demands, one has to avoid, however, making another supposition, namely, assuming an evaluative dichotomy of technocratic and 'other' approaches towards nature, as it was often the case in early studies about 'local knowledge'.[24] Non-technocratic interactions with nature may well include functionalist approaches, and vernacular knowledge is not per se holistic, harmonious and modest.[25] What is fundamentally decisive, however, is whether man's position towards nature is seen as external or as part of the surrounding environment he lives in, in other words, whether the relationship to the non-human is unilateral or reciprocal. The respective image of nature determines the recognition of water bodies and has decisive impacts on the role a nation-state ascribes to them. When rivers are seen as exhaustible resources, their reorganisation and redistribution in the name of the well-being of society appear consequently justified. Dams and other hydropower projects irreversibly change usage patterns of rivers and ascertain – post-implementation – access to rivers and reservoirs.

[23]Makropoulos refers here to Blumenberg's usage and interpretation of the term 'techné'.

[24]See e.g. Brokensha et al. (1980) and their 'development from below' approach or publications like *Farmer first* (Chambers et al. 1989).

[25]It is equally important to emphasise the social distribution of knowledge. 'Local knowledge' is often euphemistically assumed to be socially more or less evenly distributed. However, a certain knowledge implies a certain social position, knowledge communities function and reproduce themselves via inclusion and exclusion mechanisms, knowledge on the one side often lives on ignorance on the other and so forth.

7.3 Imagining Waterscapes[26]

> Landscapes are given meaning, shape and form with reference to people. For the anti-dam activists, sublime dignity lies in the people's contact with and appreciation of the environment (Sharma 2002, n. p.).

Rivers are a symbol of life – everywhere in the world. They provide water for drinking and irrigation, habitats for a variety of species; in short, they nurture the land and its people. It would be inchoate to assume that the 'West' knows only technocratic relationships with nature. Art and literature are as full with allegories regarding the beauty of flowing water bodies as is a landscape aesthetic that connects images of untouched wilderness with ideas of nature-based recreation. Yet without suggesting the revival of a neo-orientalist East-West dichotomy, it can be fairly stated that in various places outside the 'Western world', the picture is even more complex. Rivers are seen as inseparably intertwined with the cultural and spiritual practice of the people living with them. Rivers are not merely seen as a source of livelihood and nourishment but also as an essential part of many cultural practices. More importantly, the relationship is not automatically seen as one-dimensional, but as reciprocal: Where rivers are respected as living beings, even worshipped as Goddesses, there should be, one could assume, no place for their crude exploitation as natural resources.[27]

In India in particular, the valuation of rivers is closely connected to the notion of sacredness. Amongst Hindus throughout the country, the river Ganges, worshipped as Goddess Ganga, is seen as the holiest of rivers, in which believers wish to have their ashes spread after death. Politically, the conception of sacredness might raise problematic questions, as the protection of Ganga tends to be seen as exclusively Hindu – as does therefore the associated landscape.[28] This 'politics of sacredness' create an idea of nature which resists technocratic control and implements a certain pathos of 'engaging with the modern'. An interesting point is the observation that arguments with reference to the transcendent can be discursively used for two very

---

[26]The term has e.g. been employed by Swyngedouw (1999).

[27]The ongoing contamination of the Ganges speaks another language, though the picture is far more complex as it is e.g. believed that Ganga has certain self-cleaning capacities. See e.g. Alley (2002).

[28]Quite recently, an ongoing debate on the linkages and intertwinements between socioecological, political and religious argumentations has evolved in the field of watershed development (see Mawdsley 2005; Sharma 2002, 2009). The political conclusions Kothari draws considering the 'blind spot' of the political left with regard to 'spiritual types of arguments' are quite thought-provoking: 'I think that one of the mistakes that a part of India's Left has made in discounting the value of religion in Indian political life is that they have allowed the Hindu right to appropriate the space that belongs in all its plurality to the people. Which is that, if the spiritual significant [sic!] of the river is important for the local community, then we must give space to the power of that connection. We mustn't allow it to be appropriated by one particular political force. It is important to sustain the fact that an extremely liberative politics can be shaped that is rooted in a spiritual context' (SID 2007, n. p.).

disparate purposes: to respond to (local) cultural belief systems – or to appropriate their language. There is a clear distinction between the rationale of an argument and its strategic utilisation. Thus, embedded, explicitly local notions of 'living with the river' and the surrounding landscape have to be distinguished from imageries constructed to serve the wider canvas of a political agenda.

Considered a sacred entity or not, the idea of landscape turns into a vital and accessible environment through its (imaginary) visualisation. This points at a crucial aspect mentioned earlier: the aesthetical component of environment – and environment planning – which is not only reflected in infrastructural approaches but also in the day-to-day life experiences of the people confronted with environmental transformation. The image that convincingly reflects those ideas is that of the flowing river: 'It is the dharma of a river to keep flowing' (Kashinath Trivedi, quoted from Baviskar 1995). With reference to large dams, the implication within the according belief system is obvious: A dammed river is a river that lost its dharma, its purpose. Therefore, damming a river, in particular damming Ganga, is supposed to have far-reaching consequences, not only for the people affected by it but also for the river itself and its (spiritual) qualities.[29] Besides, ritual practices that depend on continuous water flows are affected when rivers are dammed, led through tunnels or carry less water.[30] Yet a diminished flow has substantial impacts beyond spiritual realms. The sheer recognition of rivers as ecological entities that are part of an unpredictable nature suggests that the unlimited attempt to control their course has severe impacts. While the reduction of a torrent river to a meagre stream appears at least pitiable, its damming implies factual consequences that are quite often, as history has shown, far beyond human (hence national) control.[31]

Submergence, on the other hand, does not only affect ecological but also cultural landscapes. Once places are submerged and people relocated, cultural practices and local knowledge systems will necessarily undergo a change as well. One of the crucial aspects with regard to submergence and subsequent displacement seems to be the experience of an altering human-nature relationship. The loss of proximity that culminates in a feeling of alienation is probably the most significant aspect: The process of dislocation might actually be described as a shift from a place to a 'non-place' (Augé 1992); the transformation of landscape changes the experience of place itself. To further illustrate the propositions put forth up to now, I will briefly discuss

[29] In her study on the pollution of the Ganges, Alley describes how the concept of purity in Hindu mythology is closely connected with the idea of flow. Yet taking into consideration engineering accomplishments in transforming/shaping the river, she seems to suggest the 'free-flow image' to be an anachronistic view of the river (see Alley 2002, pp. 106f).

[30] This can be observed, e.g. in places like Uttarkashi (state of Uttarakhand, India) that are directly affected by upstream dams.

[31] The disastrous floods that followed the breach of embankments at the Kosi river in Nepal in 2008 are just one relatively recent example.

the case of the North Indian Tehri Dam. Its specific history, geographic location and the ensuing resistance in the region underline the complexity of arguments presented so far.

7.4 Mapping the Case Study

> The opposition to the Tehri Dam was not just a simple disagreement on the viability-non viability, costs-benefits, displacement-resettlement of a big project. It struck right at the heart of philosophical, cultural, religious, political and moral debates around contemporary developmental efforts (Sharma 2009, p. 41).

Though the debate has not received the public attention the *Narmada* case gained,[32] especially on an international scale, a lot has been written on the Tehri Dam. Taking a closer look at the textual sources, one finds that the corpus consists of three main narratives. Cost-benefit analyses constitute a major part of the existing literature.[33] In the same range are a large number of contributions that deal with the Tehri Dam against the background of the ecological specifications of its location. Looking at the Himalayas as a geological structure in transition, the seismological condition of the region around the dam site is discussed as a major issue of concern (see e.g. Govardhan 1993, pp. 171ff).

A third and possibly the largest body of writing comprises contributions to the field of environmental history. Dealing mainly with the anti-dam movement, they trace its roots back to the history of socioecological movements in the region, most notably the *Chipko Andolan*.[34] Other accounts look at the impact of the movement and draw comparisons between the resistance against the Tehri Dam and other agitations, in particular the *Narmada Bachao Andolan*. Whereas one of the aspects of differentiation highlights the unlike caste and class set-up, another emphasises that in the case of the Tehri Dam agitation, 'spiritual' arguments have played a major role: 'Tehri Dam became a means of combining sacredness with impulse, gravity of high politics with solemnity of daily worship, and nature with nationalism. It transformed a river into a political, moral and emotional idea' (Sharma 2009, p. 41). Negative consequences are associated with this damming project, also because two

---

[32]The *Narmada Bachao Andolan* has decisively shaped the discussion on resistance against dam building in India and worldwide during the last decades and thus has a great importance for any discussion on dams, resistance and the respective notions of modernity. For an account of the Narmada valley, see the detailed, yet controversially received ethnography of Baviskar (1995). For various text sources, updates of events, press clippings and other information regarding projects in the Narmada valley and beyond, see also the website www.narmada.org.

[33]For an overview of the Tehri Dam Project, see e.g. INTACH (1987), Paranjpye (1988) and Govardhan (1993).

[34]The *Chipko Andolan* was supposedly the first environmental movement of its size in South Asia and became popular worldwide as the 'Hug the trees' movement. For its history and impact, see e.g. Dogra (1992), Guha (1989), Shiva (1991) and Linkenbach (1994).

of the most sacred cities in India, Haridwar and Rishikesh, are located just about 100 km downstream on the river bank. The problems anticipated not only include the fear of a devastating flood in case of a dam failure but also envisage water shortage for ritual bathing activities.[35]

A lot of the literature on Tehri deals with the specific role of Ganges, since the erection of a dam on the holiest of Indian rivers seems to be of special significance. Within the context of river, dam and nation, spaces for reconsiderations and reconfigurations of the discourse on water resources and its proper use within a given social system open up. As Mukul Sharma emphasises, '[...] visions of Ganga as holy-mother, as national security-unity, as fear-revenge, criss-cross other debates like costs-benefits and displacement-rehabilitation regarding the dam, and at times undermine far more complex issues of democracy, social justice and sustainable development' (Sharma 2009, p. 35).

The history of the Tehri Dam reflects the complexity of the discourse as depicted above; it illustrates the characteristics of a nationally envisaged development vision – its planning, its materialisation, its impact and its perception. Tehri Dam was devised in 1949 by the *Geological Survey of India*, but it was not until the early 1960s that investigations on the project were initiated, and in 1972 the dam was finally endorsed by the *Planning Commission*.[36] It took another 6 years until the construction work started under the guidance of the *Irrigation Department of the Government of Uttar Pradesh* (GUP). Until 1989 the project realisation was under the direction of the GUP before it was shifted to the newly established *Tehri Hydro Development Corporation* (THDC). Work on the project processed slowly in the early 1980s and its realisation seemed questionable until a major financial backup from the Soviets was agreed upon in 1986 during a visit of Mikhail Gorbachev to India.[37] A detailed project report that was submitted by the Irrigation Department in 1969 anticipated the dam with a final height of 260.5 m above ground level and an initially installed capacity of 600 MW hydroelectric power. After several revisions and the envisaged construction of another dam – Koteshwar – as well as a second powerhouse downstream, the final power generation capacity was predicted to be 2,400 MW. The reservoir was envisaged to irrigate 270,000 ha of land and supply about 500 ft^3/s of drinking water to Delhi and other cities in Uttar Pradesh. Filled up to its maximum height, the reservoir is about 42 km^2 large. It literally submerged the whole Bhagirathi valley and the town of Old Tehri, 23 villages nearby and more than 70 other villages at least partially. The number of persons displaced can be estimated to about 85,000 (Sharma 2009, p. 36).

---

[35]According to a poster at Bahuguna's camp site in 1996 that was often quoted afterwards, Rishikesh would be submerged in 63 min in case of a dam failure and Haridwar within another 17 min (quoted from Bhattacharjea 1996).

[36]For a chronology of the events, see Chronology of Controversy (2004) and IRN (2002).

[37]Various sources suggest that the Tehri Dam Project was quickly chosen to fill the 'diplomatic gap' since negotiations for an investment of Soviet financial aid in the nuclear sector had failed (see e.g. Pearce 1991, p. 124).

Ever since the Tehri Dam was drafted, it has been treated with scepticism and discredit. At various levels, governmental as well as scientific, the reputation of the project was not a particularly positive one. Criticism was raised by high-ranking politicians – Indira Gandhi herself was sceptical of the project's implementation – as well as legal authorities and seismological experts. Various assessment committees have been set up since the first planning stages of the project, mainly to investigate seismological and other ecological risks associated with its implementation. Nearly all of the reports submitted agreed that the project should not be implemented unless its design would be fundamentally corrected. Comments by foreign experts, e.g. Soviet engineers, painted a similar picture (see e.g. Paranjpye 1988, pp. 24ff and N. N. 2004). As the dam's materialisation became apparent in public, resistance arose on a larger public scale. *The Tehri Bandh Virodhi Sangharsh Samiti* (TBVSS, committee to oppose the Tehri Dam), founded by Virendra Datt Saklani, drew upon the legacy of the history of social movements in the state. A veteran from the pre-independence freedom movement like his fellow combatant Sunderlal Bahuguna, Saklani was part of the *Chipko Andolan* before he founded the TBVSS in 1978.

The struggle reached its peak in 1992 with Sunderlal Bahuguna's first hunger strike, which attracted attention and support beyond regional boundaries and led to the assurance of a project review. For the next couple of years, however, no official decision was taken neither by the state nor the central government, which might have led to a diminishing of the struggle. It is suggested that the movement was further weakened by the death of Saklani, the quitting of some of its senior activists and the fact that it thereafter relied mainly on the figure of Bahuguna (Pathak 2005, p. 3638). It is not the aim of this chapter to analyse possible failures or successes of the movement. It seems noteworthy, however, that movements teach lessons regardless of their factual impact – about the nation, the state of (civil) society, spaces for dissent and future interventions.[38]

With the completion of the Tehri Dam in 2002, the story of hydropower and its impacts in Uttarakhand has not come to an end. While the devastations in the valley are obvious, there are several other highly contested hydropower projects located throughout the state, some of which reshaped its rivers and valleys severely. Being labelled the 'water tower' of India, Uttarakhand faces the construction of about 200 new hydropower projects that are either envisioned or already under construction.[39]

To date, a lot of the regional agitations connect their legacies to *Chipko* heritage and, more generally, 'Gandhian' principles of resistance. This is of particular interest, since thoughts and practices invoked in the name of Gandhi reflect what has once been the 'other' great vision for India's post-independence continuance.

[38]For a brief historical sketch of the movement, its supposed disadvantages and the 'lessons' to be learnt for future movements, see Pathak (2005).

[39]For a detailed map of the projects envisaged and under construction in Uttarakhand, see South Asia Network on Dams, Rivers and People (SANDRP 2009).

As Ashis Nandy remarks: 'The Gandhian world view may have gone underground but it remains a powerful world view and seductively relevant. It does not allow us to talk glibly about separating religion and politics. [...] However, if one looks at the country's ruling culture of politics, one finds that the space he occupies has shrunk and survives mainly as an underground consciousness that often clashes with or defies the priorities of the Indian nation-state [...]' (Nandy 2007, pp. 51ff).

As mentioned above, dissidence in the region partakes in a complex setting between the river, the dam and the nation. Interestingly, the arguments which have been evoked to oppose the dam, the 'sanctity of science', the sacredness of Ganga or the 'balance sheet' (Pearce 1991, p. 128) have often been employed and mutually turned upside down by opponents as well as proponents of the dam. Just as the 'sanctity of science' can justify the necessity of large-scale projects and interventions, scientific scrutiny could be used as a tool to criticise technical and methodological deficiencies by opponents of big projects. The sacredness of the river evoked as a spiritual argument against its damming could be responded to by the proponent side in assuring that specific spiritual qualities of the river would not be affected. Not only have positions been adjusted by systematically including issues – such as environmental concerns – raised by the opponents' side, but the semantics employed by proponents of dams also seem to be loaded with appropriated arguments. This, ironically, turns out to be not least an effect of the increased discursive presence of dams opponents.[40] Just like resistance movements have made strategic use of tools provided by the public sphere (like the media) or the government (like the legal institutions), the lobby in favour of the relevant projects has made affirmative use of the discursive space opened up its opponents. In a globalised knowledge society – depending on the definition of knowledge, one might rather speak of an information society – mutual reproductions of knowledge and power predetermine discursive structures and influence public and scientific debates about asymmetrical relations between metropolis and periphery as well as claims for (alternative) development and its cultural and educational foundation. To achieve a substantial speaker position, it becomes crucial to hold an 'insider status': It is decisive what is inside and what is 'outside the global knowledge economy' (see Kaika 2006, p. 277) – and for what reason. This insight will help us to formulate an initial answer to the second question regarding the potential space for critical articulation raised in the introductory section of this chapter.

[40]The report *A place for dams in the 21st century?* by the World Wildlife Fund (WWF) indicates that major international institutions like the ICOLD (International Commission on Large Dams) and the World Bank are following exactly that strategy: 'Bowing to domestic and international pressures, proponents have "corrected their language" and modified their positions [...].', see WWF (2000, p. 37, pdf-document, page numbers not indicated).

7.5 Exploring Spaces

> The globalization of the theory of development [...] may serve as empirical evidence of the appeals of modernization only so long as the discourse ignores the ongoing resistance to modernization by portraying as backward any such resistance, and the deployment of power in overcoming it (Dirlik 2002, p. 20).

As suggested above, the restriction of the discursive space in which criticism can be articulated is, if nothing else, a result of the historical process of its emergence that led to a (social) constellation in which its founding principles appear to be self-evident. To understand this condition it is essential to deconstruct the power mechanisms that shape the discursive field, which predetermines the requirements an articulation must fulfil to be heard and to gain meaning within the hegemonic discourse. Yet by postponing contemporary engagements at the cost of contemplating the 'big picture', one might not only discredit commitments but also fail to respond to the 'needs of the present'. What Pearce states for the anti-Tehri Dam struggle is likely to be true for a number of resistance movements: '[...] for the majority of the people of the Bhagirathi valley it matters only that the dam be halted' (Pearce 1991, p. 128). This reflects the dilemma of any critique that responds to the pragmatism of realpolitik.[41]

Anti-dam activists, advocates of an alternative development model or simply people that question the common rules of resource distribution, are typically discredited on two grounds. They are either depicted as naïve with concepts based on unrealistic and therefore invalid economical calculations or denounced as arguing on an 'emotional' and therefore unscientific basis. Confronted with such a discrediting, quite a few would respond in a defensive matter that draws on cost-benefit analyses, statistical data and environmental impact assessments in order to prove that there is scientific substance to their argument.[42]

Nonconformist arguments are not acceptable when the urgent needs of the present are being used to justify particular sacrifices for 'the greater common good'. This aspect hints at another solemn argument that is employed against opponents of large-scale interventionist projects: They are regularly depicted as anti-national. This reflects the intricate relationship between the Indian nation-state and its critics from the moment of its formation and could partially explain the difficulties critical articulation faces. The ambiguous position of the post-colonial state, 'its rare combination of power and utter dominance over the moral imagination of its people' (Kaviraj 2003, p. 159), restricted the space for the emergence and articulation of resistance in the form of a consistent civil society. It was not least the nationalist rhetoric that created an 'illusion of consensus' in early post-independence days,

---

[41]This does not imply a valuation. It is meant to indicate that the respective criticism refers to an inequality in resource distribution and gain-loss balances, its major aim therefore being an enhanced equilibrium between social costs and benefits. Thus, it reflects a reactive or responsive type of criticism.

[42]This reflects the observation of a general tendency. There are remarkable exceptions. See e.g. Roy (2000, 2001).

which led to a (historical) constellation that made the formation of non-state activism first seem gratuitous and later unfeasible (ibid. 160). 'Any group or interest that spoke about restricting the new nation-state's power could be suspected of betrayal' (ibid. 159). The rhetoric of the greater common good – anti-colonially or neo-liberally motivated – appears therefore difficult in a dual sense: On the one hand, it determines and significantly reduces the space for the (subaltern) articulation of dissent. On the other hand, it conceals the historical contingency of so-called developmental values and naturalises the demands and norms of its self-appointed representatives. As shown in the first section of this chapter, the first step is thus to historicise paradigms and to contextualise their legacies.

Let me return to a question with which I started, as it might give an answer to the other question raised above: If it is obvious from a century-long experience that mega dams are more often than not ecologically and socially problematic, economically unproductive and outdated in their modernist vision, then why would their construction still be high on a nation's agenda?

We could resort to a metaphor at this point. Top-down developmental politics, under the range of which large dams can be reckoned, habitually function like a Rubik's Cube: One starts off with a multisited object that is all scrambled and messy, but through the application of a few moves and turns, it can be given a nice and neatly shaped appearance. One needs to know the proper algorithms though. The simplest method to solve the Rubik's Cube is to adjust it layer by layer. Yet there are several tricky aspects. Once one layer is set in order, one has to address the next one with another algorithm. There is a chance that the first one gets scrambled again while doing that, especially if one does not pay attention to what is happening at the bottom of the Cube while addressing its top. It is precisely that blind spot of not being able to focus on all sides at once – namely, lacking the ability to see three-dimensionally – that offers the solution to this dilemma. If you direct the spectators' attention to the order at the top, he will be so blinded by its appearance at first sight, that he will eventually not insist on lifting the bottom up to check if it is adjusted properly. If it should actually happen that somebody is not impressed by the appearance of the first layer and requests to see the large picture, the Cube's architect could point at the complexity of the sequence of algorithms. Somebody who is not familiar with the Cube's principles would not know that the method to adjust all layers is in fact simple, and that it is only the rush to make one side fit to be shown that causes the disarray of the other layers.

This allegory should have made one thing obvious: top-down development approaches are functional. They are functional mainly because they have succeeded in appearing non-negotiable, and they will remain functional unless the core principles of their architecture are questioned. Within the framework of a hegemonic development pattern that prefers big projects and big interventions, decisions that favour the implementation of large dams will continue to be taken – and they are reasonable as they are logical within the paradigm. To this extent Iyer is right: 'If we say 'No' to big dams, we should be prepared to say 'No' to those other things as well' (Iyer 2006, n. p.). The consequence is yet to be drawn. Practically as well as theoretically.

Given that they are part of national capacity building, post-independent water resource management practices in India have by trend favoured the greater common good at the cost of regional requirements. This discrepancy does not only exist at the level of a national-regional duality but is also replicated in a number of geographical and social disparities.[43] Conversely, 'water is not a national issue but an intensely local one' (World Bank 2005, quoted from Thakkar 2005b), which requires particularly local need classifications as well as problem-solving strategies. Participatory approaches, however, tend to fail in their purpose or even turn into being disparaging, when fundamental issues regarding the asymmetrical distribution of knowledge in social spaces are not raised. The derision of certain segments of society – including their knowledge – as 'backward' is likely to lead to a type of interventionist politics based on a strict separation between the desirable 'modern' and the replaceable 'traditional' that is declared to jeopardise national progress. A developmentalist notion of the modern implies a dichotomous approach that sets, literarily speaking, worlds apart, as it distinguishes between what is to be considered modern and what, as its counterpart, as traditional in the sense of 'antimodern'. The traditional is constructed as 'the other' in opposition to the modern, hence appears as an essentialised antagonism of development, as backward.[44] This not only negates the transformational character of tradition and its dynamic capacity of handling new, external influences and syncretising different knowledge bodies, it also comprises a number of serious consequences: What is constructed as backward does not contribute to developmental goals. Yet since national planning aims at taking hold of the periphery, the forced inclusion of the 'backward' segments in the top-defined development process gains its legitimacy. In a self-perpetuating process, defined marginalisation leads to the need for the social control of the marginalised.[45] The creation of a modern-backward dichotomy does, by the way, not only affect humans but also the way technologies are considered. It is quite enlightening what Sengupta says about irrigation and water harvesting systems of India: 'These systems have withstood the test of time. Their history of survival could have been regarded as proof of their eco-viability and efficiency. Instead, once they are labelled 'traditional', modern engineers tend to despise them, and development experts reject them. [...] Some of the old water appropriation systems were understood and adopted by the engineers and came to be known as 'modern' techniques. The rest, which did not attract them, were rejected as 'traditional'' (Sengupta 1993, pp. 9f).

---

[43]One can think of inequalities in the distribution that privilege urban over local populations, high cast and class segments of society over indigenous groups and men over women.

[44]This issue becomes relevant when rural and urban areas suddenly become neighbourhoods in the case of involuntary displacement. The relocated people are not only confronted with the difficulties of adapting to a new environment, they are also exposed to stigmatisation as 'backward' and 'primitive in their new urban locales' (Baruah 2006, p. 63).

[45]Typically, the assertion of the manifold problems of those segments of society that do not incorporate the attributes of modernity – nor the structural possibility to challenge its hegemonic paradigms – rationalises large-scale projects and interventions. Needless to say, the rhetoric and the typical output do not necessarily match, and those who serve as role models for the envisaged optimisation are quite often the last to benefit.

In this regard, works like those of Anupam Mishra (Gandhi Peace Foundation, New Delhi) are quite instructive. His in-depth studies of 'traditional' water harvesting methods in Rajasthan and elsewhere show that criticism of big projects does not necessarily have to be formulated as a negative account of project-associated lacks and deficiencies.[46] The illustration of local technology and infrastructure that has proven its functionality over time and is manageable in accordance with the knowledge of the people using it seems to offer a vital perspective for the evaluation of big projects beyond cost-benefit reflections.

References

Alley KD (2002) At the banks of the Ganga. When wastewater meets a sacred river. The University of Michigan Press, Ann Arbor

Augé M (1992) Non-lieux. Introduction à une anthropologie de la surmodernité. Seuil, Paris

Baruah S (2006) India and China: debating modernity. World Policy J 23(2):62–70

Baviskar A (1995) In the belly of the river. Tribal conflicts over development in the Narmada Valley. Oxford University Press, New Delhi

Bhattacharjea A (1996) The old man and the dam. Outlook Magazine, 26 June 1996. http://www.outlookindia.com/article.aspx?201604. Accessed 28 Mar 2010

Bizer JR (1987) Damning large dams. A review of Goldsmith, E. and Hildyard, N. (1984) The social and environmental effects of large dams. Ecology 68(3):757

Brokensha D, Warren DM, Werner O (eds) (1980) Indigenous knowledge systems and development. University Press of America, New York

Chakrabarty D (2000) Provincializing Europe: postcolonial thought and historical difference. Princeton University Press, Princeton

Chambers R, Pacey A, Thrupp LA (eds) (1989) Farmer first: farmer innovation and agricultural research. IT Publications, London

Chandrasekhar I, Seel PC (eds) (2003) Body city: siting contemporary culture in India. Tulika Books, New Delhi

Chatterjee P (1997) A possible India. Essays in political criticism. Oxford University Press, New Delhi

Chatterjee P (2007) Modernity and Indian nationalism. In: Jahanbegloo R (ed) India revisited. Conversations on continuity and change. Oxford University Press, New Delhi, pp 45–50

Chronology of Controversy (2004) Down Earth 13(20040831). http://www.indiaenvironmentportal.org.in/node/18934. Accessed 28 Mar 2010

CSE (Centre for Science and Environment) (1997) Dying wisdom. Rise, fall and potential of India's traditional water harvesting systems. CSE, New Delhi

D'Monte D (1985) Temples or tombs. Industry versus environment. Three controversies. CSE, New Delhi

D'Souza R (2006) Drowned and dammed: colonial capitalism and flood control in Eastern India. Oxford University Press, New Delhi

D'Souza R (2008) Framing India's hydraulic crisis. The politics of the modern large dam. Mon Rev 60(3):112–124

[46]See in particular Mishra (1993, 1994). For an overview of traditional water harvesting systems in India, see also CSE (1997); for various regional case studies, framed well in the form of travelogues, see Jacob (2008).

Dirlik A (2002) Modernity as history: post-revolutionary China, globalization and the question of modernity. Soc Hist 27(1):16–39
Dogra B (1992) Forests, dams, and survival in Tehri Garhwal. Bharat Dogra, New Delhi
Gispen K (2000) Untitled review of Laak, Dirk van: Weisse Elefanten: Anspruch und Scheitern technischer Grossprojekte im 20. Jahrhundert. J Mod Hist 72(4):1001–1002
Goldsmith E, Hildyard N (1984) The social and environmental effects of large dams, vol 1 & 2. Sierra Club Books, San Francisco
Govardhan V (1993) Environmental impact assessment of Tehri Dam. Ashish Publishing House, New Delhi
Guha R (1989) The unquiet woods: ecological change and peasant resistance in the Himalaya. Oxford University Press, New Delhi
Guha R (2005) Prime ministers and big dams. The Hindu, 18 Dec 2005
INTACH (Indian National Trust for Art and Cultural Heritage) (1987) The Tehri Dam: a prescription for disaster. A selection of recent writings on the Tehri Dam project. INTACH, New Delhi
IRN (International Rivers Network) (2002) Tehri fact sheet. http://www.internationalrivers.org/files/TehriFactsheet2002.pdf. Accessed 4 Mar 2007
Iyer RR (2006) Water champion: Ramaswamy R. Iyer. Caution and care in building dams. Interview with Christina Dueñas, Asian Development Bank. http://www.adb.org/Water/Champions/iyer.asp. Accessed 30 Aug 2009
Jacob N (2008) Jalyatra. Exploring India's traditional water management systems. Penguin, New Delhi
Jahanbegloo R (2007) India revisited. Conversations on continuity and change. Oxford University Press, New Delhi
Kaika M (2006) Dams as symbols of modernization: the urbanization of nature between geographical imagination and materiality. Ann Assoc Am Geogr 96(2):276–301
Kaviraj S (2003) In search of civil society. In: Chandrasekhar I, Seel PC (eds) Body city. Siting contemporary culture in India. Tulika Books, New Delhi, pp 151–169
Kaviraj S (2005) An outline of a revisionist theory of modernity. Archives Européenes Sociologie XLVI(3):497–526
Khagram S (2004) Dams and development: transnational struggles for water and power. Cornell University Press, Ithaca
Klingensmith D (2007) "One valley and a thousand": dams, nationalism and development. Oxford University Press, New Delhi
Klotz H (1999) Kunst im 20. Jahrhundert. Moderne – Postmoderne – Zweite Moderne. Beck, München
Linkenbach A (1994) Ecological movements and the critique of development: agents and interpreters. Thesis Elev 39:63–85
Makropoulos M (2000) Historische Kontingenz und soziale Optimierung. In: Bubner R, Mesch W (eds) Die Weltgeschichte – das Weltgericht? Akten des Internationalen Hegel-Kongresses 1999. Klett-Cotta, Stuttgart, pp 1–14
Makropoulos M (2004) Kontingenz. Aspekte einer theoretischen Semantik der Moderne. Archives Européenes de Sociologie 45(3):369–399
Mawdsley E (2005) The abuse of religion and ecology: the Vishva Hindu Parishad and Tehri Dam. World Views Environ Cult Relig 9(1):1–24
McCully P (1996) Silenced rivers: the ecology and politics of large dams. Zed Books, London
Mishra A (1993) Aaj Bhi Kharein Hai Talaab. Gandhi Peace Foundation, New Delhi
Mishra A (1994) Rajasthan ki Rajat Boonde. Gandhi Peace Foundation, New Delhi
N. N. (2004) Power point. Down Earth 13 (20040831). http://www.indiaenvironment-portal.org.in/node/18934. Accessed 28 Mar 2010
Nandy A (2007) Gandhi and the Indian identity. In: Jahanbegloo R (ed) India revisited. Conversations on continuity and change. Oxford University Press, New Delhi, pp 51–58
Nehru J (1958) Temples of the new age. In: Nehru J (ed) Speeches, vol III, March 1953–August 1957. Publications Division, Government of India, Calcutta, pp 1–4

Paranjpye V (1988) Evaluating the Tehri Dam: an extended cost benefit appraisal. Indian National Trust for Art and Cultural Heritage (INTACH), New Delhi
Pathak S (2005) Tehri Dam: submersion of a town, not of an idea. Econ Political Wkly 40(33):3637–3638
Pearce F (1991) Building a disaster: the monumental folly of India's Tehri Dam. Ecologist 21(3):123–128
Prasad S (2001) Towards an understanding of Gandhi's views on science. Econ Political Wkly 36(39):3721–3732
Roy A (1999) The greater common good. http://www.narmada.org/gcg/gcg.html. Accessed 1 Dec 2008
Roy A (2000) Power politics: the reincarnation of Rumpelstiltskin. Outlook Magazine, 27 Nov 2000. http://www.outlookindia.com/article.aspx?210449. Accessed 23 Jan 2009
Roy A (2001) Scimitars in the sun. Interview with N. Ram. Frontline 18(1). http://wwwfront-lineonnet.com/fl1801/18010040.htm. Accessed 23 Jan 2009
Saha MN (1955) "Flood control projects in the Second Five-Year Plan", Speech in Parliament. In: Saha MN (ed) Papers, Instalment No. VII., S. NO. 26, ff. 17–21. Nehru Memorial Museum and Library (NMML), New Delhi
Sahasrabudhey S (2002) Gandhi's challenge to modern science. The Other India Press, Mapusa
SANDRP (South Asia Network on Dams, Rivers and People) (2009) Hydro-electric projects on River Ganga in Uttarakhand (map). http://www.sandrp.in/rivers/Hydro_Electric_Projects_on_Ganga_in_Uttarakhand.jpg. Accessed 4 Mar 2010 (in Hindi)
Scott J (1998) Seeing like a state. How certain schemes to improve the human condition have failed. Yale University Press, New Haven
Sengupta N (1993) User-friendly irrigation designs. Sage, New Delhi
Sharma M (2002) Saffronising green. http://www.india-seminar.com/2002/516/516%20mukul%20sharma.htm. Accessed 3 May 2009
Sharma M (2009) Passages from nature to nationalism: Sunderlal Bahuguna and Tehri Dam opposition in Garhwal. Econ Political Wkly 44(8):35–42
Shiva V (1991) Ecology and the politics of survival. Conflicts over natural resources in India. Sage, New Delhi
SID (Society for International Development Forum) (2007) Smitu Kothari on dams, development, and the ethics of engagement. Interview by Georgina Drew. http://www.sidintnet/smitu-kothari-on-dams-development-and-the-ethics-of-engagement/. Accessed 16 Mar 2010
Singh S (1997) Taming the waters: the political economy of large dams in India. Oxford University Press, New Delhi
Swyngedouw E (1999) Modernity and hybridity: nature, regeneracionismo, and the production of the Spanish waterscape, 1890–1930. Ann Assoc Am Geogr 89(3):443–465
Thakkar H (2005a) A reality check on Bhakra. Review of "Unravelling Bhakra: assessing the temple of resurgent India", a Manthan Publication 2005. In: Infochange news & features. http://www.sandrp.in/dams (original source: http://infochangeindia.org/200506045979/Water-Resources/Books-Reports/A-Reality-Check-on-Bhakra.html). Accessed 22 June 2008
Thakkar H (2005b) The World Bank's misguided advocacy of large dams in India. In: Infochange news & features. http://www.infochangeindia.org/analysis98.jsp. Accessed 22 June 2008
WCD (World Commission on Dams) (2000) Dams and development. A new framework for decision-making. Earthscan, London
WWF (World Wildlife Fond) (2000) A place for dams in the 21st century? http://www.pandaorg/about_our_earth/about_freshwater/freshwater_resources/?3660/A-Place-for-Dams-in-the-21st-Century. Accessed 29 Mar 2010
Wilson RG (1985) Machine-age iconography in the American West: the design of Hoover Dam. Pac Hist Rev Archit Am West 54(4):463–493
Worster D (1985) Rivers of empire: water, aridity and the growth of the American West. Oxford University Press, New York

Chapter 8
The Promotion of Dams Through the Clean Development Mechanism: Between Sustainable Climate Protection and Carbon Colonialism

Alexander Erlewein

Abstract The current discussion about global warming and the possibility to reduce greenhouse gas emissions through hydropower has given a new turn to the debate about dams, resulting in the re-evaluation of this otherwise disputed technology. This trend materialises in the massive financial support that the United Nations' carbon-offsetting scheme *Clean Development Mechanism* (CDM) mobilises for the construction of new hydropower plants in developing countries. As defined in the Kyoto Protocol, CDM projects are supposed to avoid greenhouse gas emissions while simultaneously contributing to sustainable development. The objective of this chapter is to analyse to what extent carbon-offsetting dams are able to live up to this 'win-win' expectation. By identifying considerable challenges and constraints, it is argued that the capability of large hydropower projects to contribute to climate protection and to sustainable development is questionable. Given the controversial effects large dams may have on the local level, it is discussed in which respect carbon-offsetting dams constitute a form of 'carbon colonialism' that results in the exacerbation of one of the most problematic aspects of global warming: the asymmetries of causation and burden-sharing.

Keywords Hydropower • Carbon offsetting • Climate justice • India

8.1 Introduction

Since the rise of environmentalism in the 1980s, the appreciation of large dams has been impaired profoundly. Reports about the destruction of river ecosystems, the struggle of resistance movements all over the world, and the often miserable fate

A. Erlewein, Dipl.-Geogr. (✉)
Department of Geography, South Asia Institute, Heidelberg University,
Im Neuenheimer Feld 330, 69120 Heidelberg, Germany
e-mail: erlewein@sai.uni-heidelberg.de

M. Nüsser (ed.), *Large Dams in Asia*, Advances in Asian Human-Environmental Research, DOI 10.1007/978-94-007-2798-4_8,
© Springer Science+Business Media Dordrecht 2014

of dam-displaced people, which are estimated to add up to 40–80 million persons worldwide (WCD 2000a), had a strong influence on the public perception of dams: 'Dam has been a dirty word for years' (The Economist 2009). From symbols of progress, dams had evolved into controversial large-scale projects and, for some, even into examples par excellence of failed development interventions (Roy 2001). However, the debate on large dams has taken a new turn recently. Hydropower dams are experiencing a positive re-evaluation as a means of mitigating climate change. In addition to other important factors such as the strongly increasing energy demand of emerging economies or the aspiration of some countries to become more independent from volatile energy imports, it is the 'clean and green' credentials on hydropower that are leading to a renewed interest in dams: 'Hydropower continues to be the most important and economic source of commercial renewable energy worldwide, and its popularity is increasing with the surge of interest in clean energy prompted by concerns about climate change' (UNESCO 2009, p. 118). The share of hydropower in the sector of renewable energy is dominant. It accounts for around 83 % of the globally produced renewable electricity, the vast majority of the approximately 18 % that all renewables together contribute to the total electricity generation (REN21 2009). Although hydropower is an 'old-timer' in the group of renewable energies, it benefits substantially from the financial support that governments and companies mobilise in order to promote the dissemination of regenerative technologies. This is most evident through the incorporation of hydropower into the trade with emission certificates. The United Nation's *Clean Development Mechanism* (CDM), one of the most important instruments of international climate policy, recognises hydropower dams as sustainable carbon-offsetting projects and supports the construction of hundreds of large-scale projects (>15 MW) in newly industrialising and developing countries through the allotment of tradable carbon credits. However, the contribution of these carbon-offsetting dams towards fulfilling the goals of the CDM in terms of climate protection and sustainable development is disputed (Haya 2007; Nagle 2009). Moreover, the outsourcing of controversial climate protection projects to the South gives rise to questions regarding equity and justice in climate change mitigation (Lohmann 2008; Wright 2007).

Against this background, the aim of this chapter is to analyse carbon-offsetting dams with respect to (a) their ability to advance the goals of the CDM and (b) their implications for climate justice. The text is structured into three sections. The first section explores the rationale of carbon offsetting and outlines the administrative structure of the CDM as well as the most important regulations governing the accreditation and implementation of CDM projects. Subsequently, the CDM support for hydropower development is examined. By identifying various challenges and constraints, it is analysed to what extent carbon-offsetting dams are able to contribute to the CDM objectives of climate protection and sustainable development. In the third section, these findings are discussed with respect to their implications for climate justice: Under which circumstances are CDM dams in a position to mitigate the asymmetries of climate change causation and burden-sharing, and when do they constitute cases of 'carbon colonialism' that exacerbate existing disparities?

8.2 The Clean Development Mechanism

The CDM is part of the Kyoto Protocol (UNFCCC 1998), the first internationally binding climate protection treaty. This agreement is characterised by a strong reliance on market-based mitigation strategies, an approach that has become known as 'cap and trade' (Oberthür and Ott 1999). The protocol's target to reduce the greenhouse gas (GHG) emissions of industrialised countries by 5.2 % below the 1990 baseline until 2012 ('cap') is supplemented by three so-called flexible mechanisms[1] which are supposed to facilitate the implementation of reduction efforts through market-based instruments ('trade'). In addition to the trading of those emission certificates that have been assigned to industrialised countries (so-called Annex I countries) based on their national reduction targets, emission certificates, also referred to as carbon credits, can be generated by investing in climate protection projects abroad. Against this background, the CDM enables Annex I countries (or companies that face binding emission restrictions within these countries) to invest in climate protection activities in Non-Annex I countries, i.e. in newly industrialising and developing countries without binding emission reduction targets. The carbon credits generated through such a project activity, so-called Certified Emission Reductions (CERs), are transferred to the investing Annex I country and are calculated towards the country's reduction target (see Fig. 8.1).

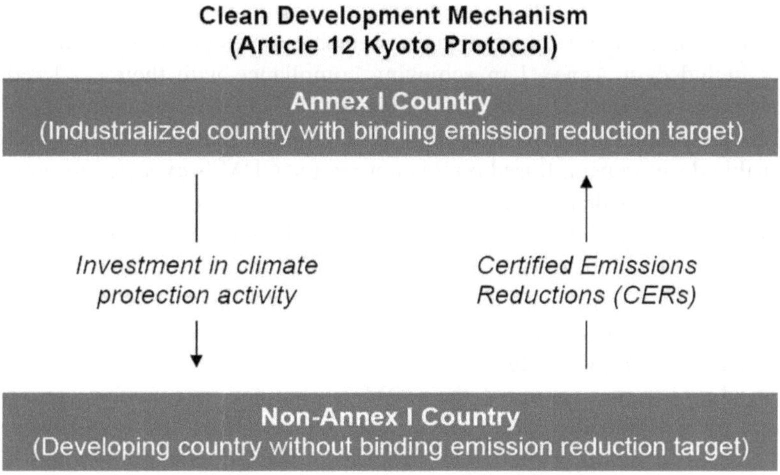

Fig. 8.1 Basic structure of the Clean Development Mechanism (*CDM*)

[1]The three 'flexible mechanisms' of the Kyoto Protocol are Emissions Trading (ET), Joint Implementation (JI) and Clean Development Mechanism (CDM).

Accordingly, the CDM is an international carbon-offsetting scheme that enables industrialised countries to compensate for excess GHG emissions by purchasing carbon credits from climate protection projects in the South.

The concept of carbon offsetting in developing countries is based on two essential ideas. First, it does not matter where GHGs are emitted or reduced. The principal GHGs stay in the atmosphere long enough to mix uniformly over the entire globe. Thus, their global warming potential is independent from the location of the emitting source (IPCC 2007a). Secondly, reducing GHG emissions in the developing world is less expensive than in the industrialised North. For example, reducing the emission intensity of an old inefficient thermal power plant in India is more economic than retrofitting an already very efficient thermal power plant in France. Consequently, the CDM takes advantage of 'spatially differentiated emission-abatement costs' (Bumpus and Liverman 2008, p. 134), i.e. the fact that the same amount of money can reduce more emissions in the South than in the North. By commodifying carbon and creating a market for the trading of emission certificates between developing and industrialised countries, the CDM does what a market mechanism is supposed to do: it aims to identify the most economic possibilities to reduce GHG emissions, so-called low-hanging fruits, in order to allocate limited resources efficiently (Oberthür and Ott 1999). Thus, in the strict sense, the CDM is not an instrument for climate protection but an instrument to reduce the costs of mitigation efforts.

The second objective of the CDM is to promote sustainable development in the host countries of CDM projects. Article 12.2 of the Kyoto Protocol states that 'The purpose of the clean development mechanism shall be to assist Parties not included in Annex I in achieving sustainable development [...] and to assist Parties included in Annex I in achieving compliance with their [...] reduction commitments' (UNFCCC 1998, p. 11). Hence, the CDM postulates a direct nexus between the implementation of climate protection projects and the promotion of sustainable development. Based on this notion, the CDM is expected to function as a 'win-win' strategy that benefits all stakeholders. Developers of CDM projects can gain income by selling carbon credits; Non-Annex I countries and local stakeholders are supposed to benefit from a project's contribution to sustainable development, while Annex I countries and their industries can meet their reduction targets more economically by purchasing inexpensive carbon credits instead of investing in more costly domestic emission reductions.

The administrative set-up of the CDM is complex and involves an array of different actors from the public and private sector. Of central importance is the CDM Executive Board of the United Nations' climate secretariat, United Nations Framework Convention on Climate Change (UNFCCC). On behalf of the parties of the Kyoto Protocol, the Executive Board sets the regulations governing the CDM project cycle and decides upon the approval of individual project proposals. Furthermore, so-called Designated Operational Entities (DOE) play a key role. DOEs are UN-accredited consulting firms that are supposed to support the decision-making of the Executive Board by providing independent project assessments. The major steps and actors of the CDM project cycle, from the project proposal to the final issuance of CERs, are outlined in Fig. 8.2.

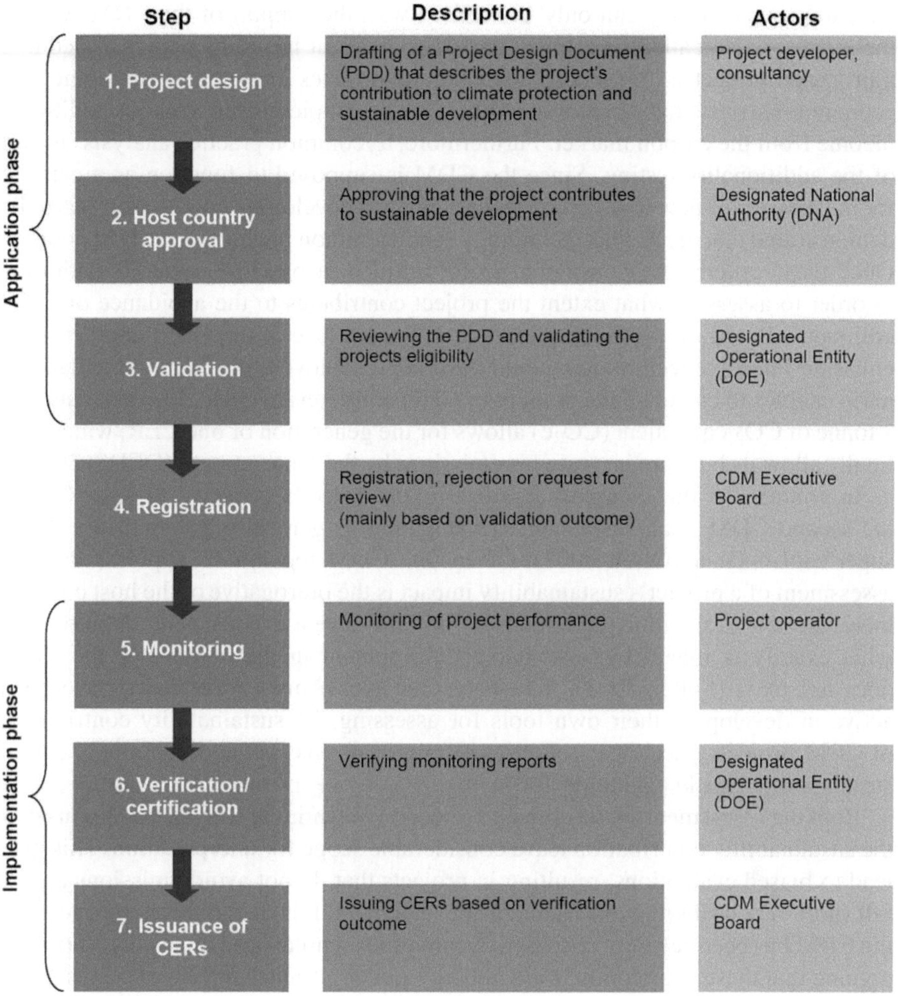

Fig. 8.2 The CDM project cycle and its actors (Adapted and extended from: Streck 2004, p. 303)

In order to ensure that a project proposal qualifies for the CDM, it is essential to thoroughly analyse to what extent the project can be expected to contribute to climate protection and sustainable development. The assessment of a project's potential to avoid GHG emissions consists basically of two procedures: the demonstration of the project's additionality and the drafting of a baseline scenario. Additionality is the elementary prerequisite for a project's contribution to climate change mitigation. Since the CDM is an offsetting scheme generating carbon credits that will be used by its buyers to maintain or increase emissions in Annex I countries, projects have to be additional to what would have happened in the absence of the CDM. Only projects that are implemented over and above the business as usual activities are able to effectively compensate emissions. Therefore, project proposals have to prove that they can only be realised with the support of the CDM. Within the scope of a so-called investment barrier analysis, it has to be demonstrated that a proposed project is financially unviable and/or faces any other technological or economic barriers that renders it unlikely to get implemented without additional income from the carbon market. Furthermore, a 'common practice analysis' is part of the additionality testing. Since the CDM is supposed to function as a catalyst for the transfer of new low-carbon technologies to developing countries, it has to be demonstrated that the project technology is not common practice in the host country. Once these criteria are considered to be fulfilled, a baseline scenario is drafted in order to assess to what extent the project contributes to the avoidance of GHG emissions. Based on the assumption that in the absence of the CDM another more emission-intensive technology would have been deployed, this counterfactual scenario enables to calculate the amount of GHG emissions avoided. The avoidance of 1 tonne of CO_2 equivalent (CO_2e) allows for the generation of one CER, which will finally allow its buyer to increase his emissions by the same amount (DEHSt 2007a).

In contrast to the complex appraisal of a project's capacity to save GHGs, the second CDM goal, sustainable development, is generally assessed in a rather superficial manner (Olsen 2007). Given that CDM regulations stipulate that the assessment of a project's sustainability impact is the prerogative of the host country, there are no evaluation guidelines or criteria. Even a consistent definition of what exactly is meant by sustainable development in the context of the CDM does not exist (Sutter 2003). Whereas some Non-Annex I countries have been active in developing their own tools for assessing the sustainability contribution of CDM projects, most host countries refrain from this task and apply only casual sustainability checks (Rudolph 2007).

Both the assessment of the climate protection potential as well as the appraisal of the sustainability contribution leave considerable scope for interpretation. This may lead to biased evaluations, resulting in projects that do not avoid emissions and/or fall short of contributing to sustainable development. It is due to this deficiency that the CDM has been severely criticised by numerous scholars and the media, variously arguing that these shortcomings are not only minor pitfalls but pose a serious threat to the environmental and social integrity of the CDM as a whole (e.g. Böhm and Dabhi 2009; Gilbertson and Reyes 2009; Lohmann 2006).

8.3 Hydropower Within the CDM

The CDM comprises a multitude of technologies that are supposed to avoid or reduce the GHGs regulated under the Kyoto Protocol.[2] Eligible are mainly renewable energy and energy efficiency projects as well as technologies that capture

[2]The Kyoto Protocol regulates six types of GHG or groups of gases: carbon dioxide, methane, nitrous oxide (N_2O), sulphur hexafluoride, hydrofluorocarbons (HFCs) and perfluorocarbons (PFCs) (UNFCCC 1998).

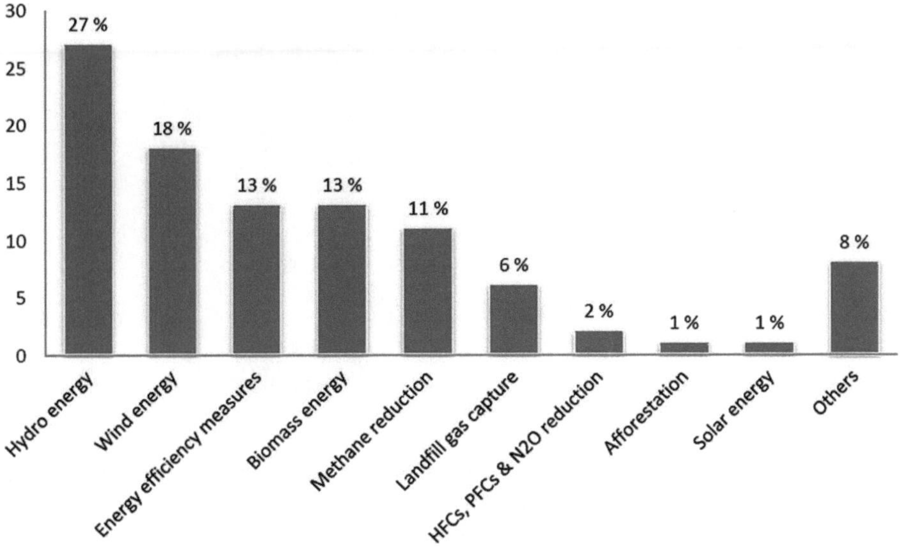

Fig. 8.3 Projects in the CDM pipeline by type as of 1 July 2010 (Data source: UNEP Risø Centre 2010)

and burn certain industrial or landfill gases. Hydropower is by far the most common project type, accounting for more than one quarter of all projects in the CDM pipeline (see Fig. 8.3).[3] In terms of expected CER generation until 2012, hydro ranks first alongside hydrofluorocarbon (HFC) reduction projects, both of which will receive around 17 % of the total amount of CERs during this period (UNEP Risø Centre 2010). Given that HFC is a particularly aggressive GHG and CERs are calculated on the basis of CO_2e, HFC reduction projects, although very small in number, generate an extraordinarily large quantity of carbon credits.

Considering the fact that the CDM is only up and running since 2004, the total number of hydropower projects that have requested CDM status is remarkable. As of 1 July 2010, there are 1,454 hydro projects in the CDM pipeline out of which 641 have already been approved by the Executive Board, whereas the remaining are in the process of application. Out of the total number, 781 hydro projects pertain to the category 'small scale' (≤15 MW) and 673 to 'large scale' (>15 MW). Altogether they represent an electricity generation capacity of 53,044 MW (UNEP Risø Centre 2010). Within only 6 years, carbon-offsetting dams gained a share of more than 5 % of the worldwide installed hydropower capacity, which is estimated to add up to almost 1,000 GW (REN21 2009). Based on the official assumption that in the absence of the CDM the generation capacity of carbon-offsetting dams would have been installed through the construction of mostly nonrenewable power plants, CDM dams are expected to avoid 492.1 million tonnes of CO_2e and generate the

[3]The CDM pipeline comprises all approved and applying projects (UNEP Risø Centre 2010).

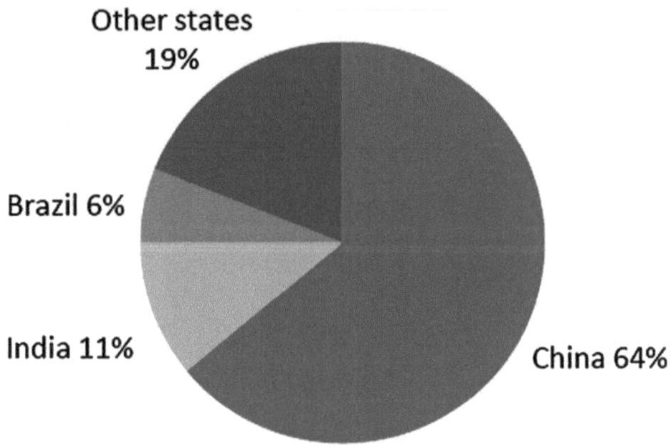

Fig. 8.4 Geographical distribution of carbon-offsetting dams as of 1 July 2010 (Data source: UNEP Risø Centre 2010)

same amount of carbon credits until 2012 (UNEP Risø Centre 2010). At the current trading value of around 13 euros per CER, these carbon credits are worth more than six billion euros, a substantial revenue stream for project owners and the associated financial institutions.

CDM hydro projects are classified as 'run of river', 'new dam', or retrofitting of 'existing dam' (UNEP Risø Centre 2010). However, clear definitions for the use of these categories in the CDM are missing. A closer examination of the technical features of CDM hydro projects reveals that the term 'run of river' is used very broadly. Even projects that allow for several days water storage or include the diversion of whole rivers are labelled as 'run of river', which is clearly contradictory to the respective river engineering definitions (Strobl and Zunic 2008). For this reason, the term 'carbon-offsetting dam' in this text also refers to projects which are officially categorised as 'run of river'.

The geographical distribution of CDM projects is highly uneven. More than 70 % of all projects are located in only three countries: China, India, and Brazil (UNEP Risø Centre 2010). In the case of carbon-offsetting dams, this disproportionate allocation is even more pronounced: 64 % are to be found in China, followed by India with a share of 11 % (see Fig. 8.4). Altogether almost 80 % of all CDM hydro projects are located in Asia, many in peripheral high mountain areas.

8.4 Analysing the Ability of Carbon-Offsetting Dams to Advance the CDM Goals

8.4.1 Climate Protection Through Carbon-Offsetting Dams?

The first question that arises with respect to the primary goal of the CDM is elementary: Is hydropower even a technology that can reduce GHG emissions? Usually hydro is considered to be an emission-free mode of electricity generation. Although this generally holds true for genuine run of river projects as well as for dams in cool climate regions, the use of hydropower in mitigating global warming is questionable in the case of reservoirs in the (sub)tropics (Fearnside 2004). Depending mainly on oxygen content and the amount of solid and dissolved organic material, these reservoirs can emit enormous quantities of methane, a GHG whose global warming potential is 25 times stronger than that of CO_2 (IPCC 2007a). Thus, shallow (sub)tropical reservoirs that have not been cleared from vegetation before flooding as well as reservoirs with a large inflow of organic material are especially susceptible to this phenomenon. Studies from dams in the Amazon basin show that their emission intensity per MW can outweigh that of fossil fuel power stations multiple times (Fearnside 2005). Although there is no general conclusion on the overall impact of reservoir emissions on global warming, these case studies demonstrate that dams are not per se climate friendly but, in some cases, may even exacerbate climate change: 'Methane squashes the green credentials of hydropower' (Giles 2006, p. 524). The CDM tries to account for reservoir emissions by excluding projects with a power density of less than 4 W per flooded m^2 and defining a standard value for discounting the GHG reduction potential of projects with a power density between 4 and 10 W/m^2 (UNFCCC 2010).

More problematic challenges arise when it comes to examining the additionality of carbon-offsetting dams. Whereas the imprecise consideration of reservoir emissions may lead to inexact allocations of carbon credits, non-additional projects undermine the *raison d'être* of the CDM. The verification of additionality constitutes one of the biggest challenges for the effectiveness of carbon-offsetting dams as well as for the CDM as a whole. To prove unambiguously that the economic barriers a project proposal is confronted with can only be overcome with the support of the CDM is often impossible. Every project faces obstacles and the rating of barriers as 'decisive' or 'non-decisive' is always debatable. Several scholars argue that such scopes of interpretation are systematically (mis)used in order to influence project validations (Lohmann 2006; Schneider 2007; Witt and Moritz 2008). A study by Haya (2007) showed that 35 % of all carbon-offsetting dams registered until 2007 have already been completed at the time of requesting CDM registration and another 54 % were planned to be completed within 1 year after registration. If the CDM had been of vital importance for the implementation of these projects, as the developers claim in their registration requests, one should assume that construction only starts after the CDM approval. The case of the Allain Duhangan project in the Indian state of Himachal Pradesh exemplifies this problem. The 192 MW hydropower project was officially approved by the Executive Board in May 2007 even though the project implementation agreement had been signed in 2001 and construction began in 2004 (UNFCCC 2007a). The fact that the construction started 3 years before the project had been registered under the CDM suggests that the additional income from the selling of carbon credits had not been taken into consideration when the financial viability of the project was assessed. However, the developers of Allain Duhangan argue that the necessary loans were only granted on the expectation that the project would be approved by the CDM, a widespread argument that can be found in many CDM application documents. But again, this raises questions. If projects really depend on income from the carbon market, are developers willing to start with the implementation of these projects before getting the final registration approval? Although the Executive Board's rate of rejection is low, such an approach is probably too risky for most developers (Haya 2007). This rather implies that the CDM constitutes in many cases an *additional income* but is not vital for the origination of *additional projects*. Or, as an official of the Allain Duhangan project put it during an expert interview: 'The project would have gone ahead regardless [...] the CDM is just another incentive'. This problem is not confined to carbon-offsetting dams but rather applies to almost all project types. Schneider (2007) estimates that up to 40 % of all registered CDM projects are likely to be non-additional and would have been implemented anyway. Without reducing any emissions whatsoever, non-additional projects generate carbon credits that enable its buyers in Annex I countries to maintain or even increase GHG emissions.

Furthermore, the additionality testing comprises a 'common practice analysis' in which project proposals have to demonstrate that they contribute to the transfer and dissemination of new low-carbon technologies. However, in the case of dams, this requirement is hard to fulfil. Unlike other technologies under the CDM, which are indeed new and uncommon in many regions, most forms of hydropower generation are technologically mature and widespread all over the world (McCully 2001). This applies particularly for countries like China and India, where around three quarters of all CDM dams are located. In both countries, hydropower development has been a priority on the national agenda for more than half a century. It is paradoxical that 'China has built almost half of the world's estimated 45,000 large dams and remains one of the most active dam building countries today' (WCD 2000b, p. 1) and, at the same time, argues that dams under the CDM provide new technological impulses that would not have been realised without support from the carbon market. This inevitably leads to the assumption that the common practice analysis and the respective validation are not taken very seriously. The Allain Duhangan project confirms this concern. In order to demonstrate the project's contribution to technology transfer, the responsible Designated Operational Entity (DOE), Det Norske Veritas, states in their validation report 'that at the time of project implementation agreement in 2001 there were only two large hydropower projects in north India with installed capacity of 448 and 86 MW' (DNV 2007, p. 6). Yet, the Bhakra project alone, completed in 1963 and also located in the state of Himachal Pradesh, has a capacity of 1,325 MW (Bhatia and Malik 2008). The overall hydropower capacity of the north Indian grid even amounted to 8,332 MW in 2001 (NRPC 2009). Consequently, the Allain Duhangan project cannot be said to contribute to the transfer of new regenerative technologies.

8.4.2 Sustainable Development Through Carbon-Offsetting Dams?

Given that the second objective of the CDM consists in the advancement of sustainable development, the admission of dams as CDM projects expresses the Executive Board's point of view that carbon-offsetting dams are generally appropriate for the achievement of this goal. First of all, this raises the question of what exactly is meant by sustainable development. Both terms, sustainability and development, are controversially discussed concepts that carry very different and sometimes contradictory meanings (Rudolph 2007). In order to make use of these broad concepts, context-specific explanations are unavoidable. Yet, the CDM does not offer any consistent definition that would clarify the meaning of sustainable development in the context of climate protection projects, not to mention any criteria or indicators that would allow a coherent assessment. The only stated constraint is that GHG emission reductions themselves are not considered to be sufficient for meeting the goal of sustainable development (Sutter 2003). Apart of that, the appraisal of a project's contribution to sustainable development is entirely up to the priorities of the host countries (Olsen 2007).

However, even without a CDM-specific definition of sustainable development, the assumption that dams are generally appropriate for the achievement of this goal can be questioned. As has been mentioned at the beginning of this chapter, the pros and cons of dam building have evolved into a long-standing debate which shows that dams are far from being uncontroversial (Nüsser 2003). Striking arguments on both sides of the debate make it difficult to draw a general conclusion on this issue. By reiterating the advantages of dams in terms of electricity generation, water supply and/or flood control, dam proponents depict dams as a kind of panacea for large-scale development (e.g. IBRD 2009; Turpin 2008). In contrast, dam opponents usually refer to the local level and emphasise the adverse social and environmental consequences that may result from inundations and river diversions, such as displacement, water quality degradation and destruction of wetlands and fishing grounds (e.g. McCully 2001; Roy 2001). Given that dams have different impacts on different spatial scales, assessments of the development contribution of dams have to address the distribution of costs and benefits between and among the various place-based and non-place-based stakeholder groups. Studies focusing on these distributional implications point out that the costs and benefits of large dams are often highly unequally shared (Dwivedi 2006). In many cases, the electricity and/or water provided by dams are directly transferred to urban centres, whereas marginalised population groups up- and downstream of the dam site lose access to vital livelihood assets such as land and water (Heming et al. 2001). The most comprehensive evaluation of the development impact of large dams was performed by the World Commission on Dams (WCD). After carrying out extensive country studies and surveying more than 120 dams all over the world, the WCD drew the following conclusion: 'Dams have made an important and significant contribution to human development, and the benefits derived from them have been considerable. In too many cases an unacceptable and often unnecessary price has been paid to secure those benefits, especially in social and environmental terms, by people displaced, by communities downstream, by taxpayers and by the natural environment' (WCD 2000a, p. 310). In the light of such critical findings, it is difficult to understand why dams under the CDM are considered to be generally supportive for the advancement of sustainable development. This concern is shared by the European Union, which demands that carbon-offsetting dams with a capacity greater than 20 MW must comply with WCD sustainability recommendations, if their CERs are to be traded within the Union (DEHSt 2007b). Moreover, the Gold Standard Foundation, a non-profit organisation that assigns a quality label to sustainable carbon-offsetting projects, largely excludes large hydro projects (Gold Standard 2009). Examples like the Allain Duhangan project show that such concerns are justified. This dam is opposed by large parts of the local population because it interferes directly with the water supply of downstream villages. Besides, the developers of Allain Duhangan did not adhere to environmental regulations and have been officially fined because of violating India's forest conservation act. With reference to this project and other similar cases, Thakkar notes that 'the trouble with CDM projects in India is that the government only sees them as free gifts and is not bothered if the projects indeed qualify to be a CDM project from climate point of view, from environment impacts point of view, from social impacts point of view, from technology transfer point of view or from equity point of view' (Thakkar 2009, p. 42).

Next to these technology-specific concerns, the structure of the CDM itself may hinder the achievement of the sustainability goal. Ever since the CDM's inception, scholars have pointed out that the one-sided calculation of carbon credits is problematic in this regard (Muller 2007; Sutter 2003). The amount of CERs a project receives is calculated exclusively on the basis of GHG avoidance: one ton of avoided CO_2e yields one CER. To what extent the project also fulfils the sustainability objective does not have any influence on the amount of issued CERs. Consequently, there is no financial incentive for project developers to design and operate their projects in a way that fosters sustainable development. The absence of a financial incentive can even result in a trade-off between climate protection and development contribution (Sutter and Parreño 2007). On the one hand, the CDM as a market mechanism is supposed to identify the most economic options for GHG reduction. On the other hand, the realisation of the sustainability objective usually requires additional investment, thus rendering sustainable projects more expensive. This may result in a situation in which the sustainability contribution of a project turns out to be a competitive disadvantage that undermines the objective of sustainable development in the long run: 'Competition among non-Annex 1 parties in attracting CDM investments may therefore create an incentive to set low sustainability standards in order to yield more projects with low abatement costs. This could lead to a "race to the bottom" in terms of sustainable development standards' (Sutter 2003, p. 68). This trade-off disadvantages small-scale projects in particular, as they generate only a small amount of carbon credits but may have a positive impact on local development. In summary, it can be argued that the one-sided calculation of CERs is a structural obstacle for the achievement of the sustainability goal. Indeed, the initial expectation that the CDM can succeed as a 'win-win' instrument that delivers on both the objectives of climate *and* development policy has largely vanished (Olsen 2007; Paulsson 2009). Many people active in the carbon trading business agree that the CDM today is almost exclusively considered to be a mechanism for the generation of inexpensive carbon credits. This viewpoint is further evident in the very small share of projects that seek to conform to the sustainability criteria of the Gold Standard Foundation.

Table 8.1 Main factors challenging the contribution of large-scale carbon-offsetting dams to climate protection and sustainable development

CDM objective 1: climate protection	CDM objective 2: sustainable development
Achievement through carbon-offsetting-dams questionable because of:	Achievement through carbon-offsetting-dams questionable because of:
1. Doubts about additionality	1. Partly negative experience with the sustainability impact of large dams
2. Reservoir emissions	2. Absence of financial incentives
	3. Absence of CDM specific definition of sustainable development

The previous sections have shown that several factors constrain the ability of carbon-offsetting dams to realise the CDM objectives (see Table 8.1). On the one hand, there are fundamental concerns which question the ability of dams to contribute to climate protection and sustainable development. The problem of reservoir emissions and the various negative experiences with the development impact of large dams suggest that this technology is only partially appropriate to achieve the goals of the CDM in the first place. In addition, the CDM governance structure comprises regulations that further obstruct the effectiveness of carbon-offsetting dams. Shortcomings in demonstrating additionality and the absence of financial incentives for sustainable development challenge the environmental and social integrity of the CDM. Not only may these deficiencies render CDM projects ineffective, but they may in fact result in projects that are counterproductive in terms of climate protection and sustainable development.

8.5 CDM Dams in the Context of Climate Justice: A Case of Carbon Colonialism?

The CDM as an instrument of international climate and development policy establishes new trading connections between industrialised and developing countries: 'carbon emissions are emerging as a new and dynamic commodity that links the global North and South' (Bumpus and Liverman 2008, p. 128). Thus, the modalities and circumstances under which carbon offsets are generated and traded have implications for the relationship between industrialised and developing countries and the respective disparities (Wright 2007). Against this background, the conclusion of the previous analysis on carbon-offsetting dams gives rise to questions about equity and justice in climate change mitigation.

The discussion on climate justice forms part of the broader debate on environmental justice and focuses largely on the distributional implications of climate change (Harris 2009). Given that the underlying normative concepts and ideas vary, there is no general definition of climate justice. However, there are two core arguments that are characteristic for most interpretations. The first one relates to the unequal causation of anthropogenic climate change, as 'the largest share of historical and current global emissions of greenhouse gases has originated in developed countries' (UNFCCC 1992, p. 1). Although newly industrialising countries contribute increasingly to global warming, per capita emissions in the global North will continue to greatly exceed those of developing countries (IPCC 2007a). The second core argument aims at exposing the unequal distribution of adverse climate change impacts. It is widely accepted that the negative consequences of global warming are to be felt most strongly in the global South. Many developing countries lack the financial, technological and organisational capacities to adapt successfully to the expected challenges of climate change and are therefore more vulnerable (IPCC 2007b). Thus, the causal factors of anthropogenic climate change *and* its adverse impacts are distributed in a highly unequal manner: those who hardly contributed to climate change are particularly severely affected by its negative consequences. Or, as Prouty puts it: 'The distribution of the burdens of climate change is directly inverse to the benefits attained through CO_2 emissions' (Prouty 2009, p. 517). In the context of these asymmetries, climate change mitigation inevitably involves questions of justice and accountability. This notion is well established in international climate policy and gets reflected in the principle of 'common but differentiated responsibilities' (UNFCCC 1992, p. 4). By stipulating 'that the developed country Parties should take the lead in combating climate change and the adverse effects thereof' (UNFCCC 1992, p. 4), this principle acknowledges the historic responsibility of the industrialised world and creates the legal basis for the distinction between Annex I countries with binding reduction targets and Non-Annex I countries without such obligations.

Whether the outsourcing of climate protection measures to the South is in line with this principle or not is the primary point of contention in the controversy about carbon offsetting and climate justice. Many scholars and activists maintain that offsetting enables emitters to circumvent their own GHG reductions and consequently distracts from the root cause of climate change, i.e. the overconsumption of fossil fuels in the global North (Gilbertson and Reyes 2009; Lohmann 2006). In this sense, offsetting is equated with the selling of indulgences, thereby undermining the historic responsibility of industrialised countries (Smith 2007). Basically, this critique of offsetting is a critique of the underlying polluter pays principle.[4] That is to say, it is not enough that the polluter pays, the polluter also has to change (Altvater and Brunnengräber 2008). Some authors extend this critique and argue that offsetting not only enables the continuation of pollution but directly disadvantages developing countries. Smith describes the CDM 'as a form of carbon colonialism, whereby resources of countries in the Majority World [...] are used in order to maintain the levels of material privilege (in this case, high levels of energy consumption) enjoyed by Northern countries' (Smith 2007, p. 25). From this point of view, carbon offsetting is seen as a new mode of exploitation that exacerbates the

[4]The polluter pays principle is constitutive for many environmental laws. It stipulates that the costs of pollution and related mitigation efforts have to be borne by the polluter (Bugge 1996).

existing North–South divide, or, as Bachram argues, 'emissions trading therefore becomes an instrument by means of which the current world order, built and founded on a history of colonialism, wields a new kind of carbon colonialism' (Bachram 2004, p. 15). In direct contrast to this fundamental critique, proponents of carbon offsetting point out that the CDM mobilises substantial financial transfers to developing countries and consequently meets the demands of those who call for more support for the global South (Mittendorf 2004). Even though the distribution of CDM projects is highly unequal and excludes most of the world's poorest nations, this argument is hard to dismiss. The CDM does encourage investments in developing countries and has become an important factor in the foreign trade balance of various Non-Annex I countries. Thus, from an international perspective, the allegation of carbon colonialism does not seem to hold true. Although it is undeniable that the CDM as a compensation scheme does not address the root causes of climate change, carbon offsetting does not necessarily take place at the expense of developing countries. However, an analysis that focuses only on the exchange between nation states obstructs the view of the local impacts of the CDM. Consequently, the due consideration of local level impacts leads to a different assessment of the CDM. If there are serious doubts about the effectiveness of CDM projects or, as the analysis of carbon-offsetting dams has shown, reasons which suggest that projects might even be counterproductive for the achievement of the CDM objectives, the claim of carbon colonialism can be substantiated in two respects.

Offsetting projects that cause local socio-economic or ecological problems obviously confirm the assumptions of carbon colonialism. Such projects indeed externalise the adverse side effects of certain climate protection measures to those who barely emit GHGs. Whereas the economic and political elites in developing countries profit from the selling of carbon credits, it is overall the existing marginalised groups who are affected by the adverse impacts of such projects (Smith 2007). This can be observed when peasants in peripheral Chinese mountain areas get displaced for the construction of carbon-offsetting dams (UNFCCC 2007b) or when traditional land using rights are ignored for the creation of afforestation plantations as happened in Uganda and elsewhere (Eraker 2000). Other examples include biomass projects that may take place at the expense of food production (WWF 2007) or waste incineration and landfill gas projects that conflict directly with the informal recycling market and undermine the livelihood of the urban poor (Tangri 2003). These examples show that certain offsetting activities can be socially and/or ecologically harmful and may result in the exacerbation of existing inequalities: 'On a local level, longstanding exploitative relationships and processes are being reinvigorated by emissions trading' (Bachram 2004, p. 8).

In the case of non-additional projects that fail to contribute to climate protection, the critique of carbon colonialism can be confirmed indirectly. Without avoiding any emissions, projects like Allain Duhangan generate 'hot air credits' that enable polluters in Annex I countries to increase emissions. Thereby, non-additional projects lead to the intensification of global warming. Given that the adverse impacts thereof affect vulnerable societies in the South in particular, one may argue that

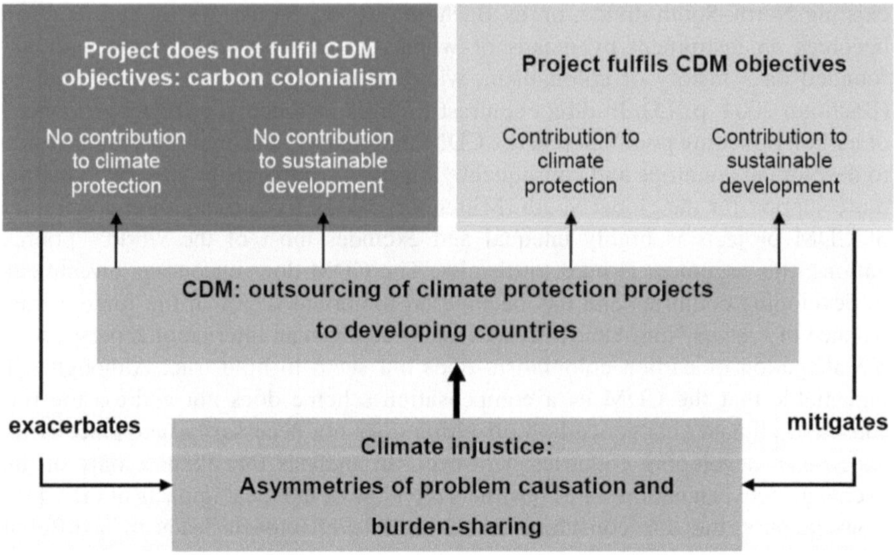

Fig. 8.5 Ambivalent consequences of the CDM on climate justice

non-additional projects eventually have an adverse effect on developing countries. In this regard, the CDM may enable the continuation of the cost externalisation that it is supposed to counteract.

As a result, it can be said that the charge of carbon colonialism does not apply to the CDM as a whole. Even though the outsourcing of climate protection measures to the South is problematic as it does not address the root causes of global warming, offsetting is not necessarily to the detriment of developing countries. On the contrary, if projects fulfil the CDM objectives, they do contribute to the mitigation of the existing inequalities of climate change causation and burden-sharing. However, evidence suggests that a significant number of CDM projects, particularly in the hydro sector, do not achieve the goals of the CDM and lead to counterproductive outcomes (Haya 2007; Schneider 2007). By increasing GHG emissions and/or causing local problems, such projects constitute a form of carbon colonialism that leads to the exacerbation of climate injustice (see Fig. 8.5).

8.6 Conclusion and Perspectives

The aim of this chapter was to analyse the ability of carbon-offsetting dams to advance the CDM goals and to assess their implications for climate justice. Building on a brief explanation of the CDM governance structure, various factors have been identified that substantially constrain the capacity of carbon-offsetting dams to live up to the 'win-win' expectation of the CDM. By discussing these shortcomings in the context of climate justice, it has finally been argued that projects which do not fulfil the CDM objectives may exacerbate the existing inequalities of climate change causation and burden-sharing.

In conclusion, this chapter points to the case that the promotion of dams through the CDM in its current form is a highly ambivalent strategy. The central problem is that carbon-offsetting dams which fail to fulfil the CDM objectives are not only ineffective but in many cases counterproductive. Given that hydropower is by far the most common CDM project type, this ambivalence seriously challenges the environmental and social integrity of the CDM. The CDM runs the risk that its worthwhile intentions and the achievements of successful projects are undermined, or *offset*, by noncompliant projects that misuse the CDM and take advantage of insufficient scrutiny procedures. If the credibility of the CDM as a meaningful instrument of international climate and development policy is to be sustained, it has to be ensured that non-additional and unsustainable projects are excluded from the CDM as far as possible. With regard to carbon-offsetting dams, there are various reform proposals which might improve the performance of the CDM.

The simplest way to avoid the risk of funding harmful carbon-offsetting dams is to exclude large-scale dams from the CDM. This is demanded by NGOs like CDM Watch, International Rivers, and SANDRP (South Asian Network on Dams, Rivers and People) who make a point in arguing that there is no reason to assume that large carbon-offsetting dams are more sustainable than ordinary large dams as long as there are no respective regulations (Haya 2007; Thakkar 2009). This leads to another reform proposal: the introduction of binding sustainability standards, such as the European Union regulation which requires large carbon-offsetting dams to comply with the sustainability guidelines of the WCD. Although the implementation of this provision is still incoherent, it provides a systematic approach for tackling some of the most severe problems associated with carbon-offsetting dams. By expanding this regulation to the whole CDM, it could be widely ensured that the CDM does not support harmful hydropower projects. Other proposals aim at weakening the position of hydro in the CDM by making carbon-offsetting dams financially less attractive for investors. Nagle (2009) proposes to discount CERs from Chinese carbon-offsetting dams in order to redirect CDM investments to countries that are in greater need of foreign support.

However, such reform proposals will remain piecemeal unless the general shortcomings of the CDM are addressed. As long as the CDM is based on hardly verifiable additionality criteria and unclear sustainability aims, it will continue to fund ineffective or, as this analysis has shown, potentially counterproductive projects which tie up financial resources that could be invested in more climate friendly and more sustainable technologies.

Acknowledgements The author gratefully acknowledges the financial support of the Cluster of Excellence 'Asia and Europe in a Global Context: Shifting Asymmetries of Cultural Flows' at the University of Heidelberg. The author would also like to thank the members of the Geography Department at the South Asia Institute of Heidelberg University as well as one anonymous reviewer for their valuable comments which significantly improved this chapter.

References

Altvater E, Brunnengräber A (eds) (2008) Ablasshandel gegen Klimawandel? Markt-basierte Instrumente in der globalen Klimapolitik und ihre Alternativen. VSA-Verlag, Hamburg

Bachram H (2004) Climate fraud and carbon colonialism: the new trade in greenhouse gases. Capital Nat Social 15(4):1–16

Bhatia R, Malik RPS (2008) Bhakra multipurpose dam system, India. In: Bhatia R, Cestti R, Malik RPS, Scatasta M (eds) Indirect economic impacts of dams: case studies from India, Egypt and Brazil. Academic Foundation, New Delhi, pp 133–193

Böhm S, Dabhi S (eds) (2009) Upsetting the offset. The political economy of carbon markets. MayFlyBooks, London

Bugge HC (1996) The principles of polluter pays in economics and law. In: Eide E, van der Bergh R (eds) Law and economics of the environment. Juridisk Forlag, Oslo, pp 77–95

Bumpus AG, Liverman DM (2008) Accumulation by decarbonization and the governance of carbon offsets. Econ Geogr 84(2):127–155

DEHSt (Deutsche Emissionshandelsstelle) (2007a) Deutsches CDM-Handbuch. Leitfaden für Antragssteller (Version 1.0), Berlin

DEHSt (Deutsche Emissionshandelsstelle) (2007b) JI und CDM: Wasserkraft-Projekte über 20 MW, Berlin

DNV (Det Norske Veritas) (2007) Validation report. Allain Duhangan Hydropower Project (ADHP) in India. http://cdm.unfccc.int/UserManagement/FileStorage/V6JOSJAHYQG5Z3U7K2OD6BMWC5G4MQ. Accessed 10 July 2010

Dwivedi R (2006) Conflict and collective action: the Sardar Sarovar project in India. Routledge, London

Eraker H (2000) CO_2lonialism – Norwegian tree plantations, carbon credits and land conflicts in Uganda. Norwatch, Oslo

Fearnside PM (2004) Greenhouse gas emissions from hydroelectric dams: controversies provide a springboard for rethinking a supposedly 'clean' energy source. Clim Chang 6(6):1–8

Fearnside PM (2005) Do hydroelectric dams mitigate global warming? The case of Brazil's Curuá-Una Dam. Mitig Adapt Strateg Glob Chang 10(4):675–691

Gilbertson T, Reyes O (2009) Carbon trading. How it works and why it fails, Critical currents 7. Dag Hammarskjöld Foundation, Uppsala

Giles J (2006) Methane squashes green credentials on hydropower. Nature 444:524–525

Gold Standard (2009) Annex C. Guidance on project type eligibility. http://www.cdmgoldstandard.org/fileadmin/editors/files/6_GS_technical_docs/GSv2.1/Annex_C.pdf. Accessed 10 July 2010

Harris PG (2009) World ethics and climate change: from international to global justice. Edinburgh University Press, Edinburgh

Haya B (2007) Failed mechanism. How the CDM is subsidizing hydro developers and harming the Kyoto Protocol. International Rivers Network (IRN), Berkeley

Heming L, Waley P, Rees P (2001) Reservoir resettlement in China: past experience and the Three Gorges Dam. Geogr J 167(3):195–212

IBRD (International Bank for Reconstruction and Development) (2009) Directions in hydropower, Washington, DC

IPCC (Intergovernmental Panel on Climate Change) (2007a) Climate change 2007: the physical science basis. Contribution of Working Group I to the fourth assessment report of the Intergovernmental Panel on Climate Change. Cambridge University Press, Cambridge

IPCC (Intergovernmental Panel on Climate Change) (2007b) Climate change 2007: impacts, adaptation and vulnerability. Contribution of Working Group II to the fourth assessment report of the Intergovernmental Panel on Climate Change. Cambridge University Press, Cambridge

Lohmann L (2006) Carbon trading: a critical conversation on climate change, privatisation and power. Dev Dialogue 48. Dag Hammarskjöld Foundation, Uppsala

Lohmann L (2008) Carbon trading, climate justice and the production of ignorance: ten examples. Development 51(3):359–365

McCully P (2001) Silenced rivers: the ecology and politics of large dams, Enlarged and updated edition. Zed Books, London.

Mittendorf M (2004) Ökonomie der internationalen Klimapolitik. Die besondere Herausforderung durch den Clean Development Mechanism. LIT Verlag, Münster

Muller A (2007) How to make the clean development mechanism sustainable – the potential of rent extraction. Energy Policy 35(6):3203–3212

Nagle JC (2009) Discounting China's CDM dams, Legal studies research paper 09-26. Notre Dame Law School, Notre Dame

NRPC (Northern Region Power Committee) (2009) Annual report 2008–09. http://www.nrpc.gov.in/Reports/ar08-09/AR2008-09.pdf. Accessed 10 July 2010

Nüsser M (2003) Political ecology of large dams: a critical review. Petermanns Geogr Mitteilungen 147(1):20–27

Oberthür S, Ott H (1999) The Kyoto Protocol: international climate policy for the 21st century. Springer, Berlin

Olsen KH (2007) The clean development mechanism's contribution to sustainable development: a review of the literature. Clim Chang 84(1):59–73

Paulsson E (2009) A review of the CDM literature: from fine-tuning to critical scrutiny? Int Environ Agreem 9(1):63–80

Prouty AE (2009) The clean development mechanism and its implications for climate justice. Columbia J Environ Law 34(2):513–540

REN21 (Renewable Energy Policy Network for the 21st Century) (2009) Renewables global status report 2009 update, Paris

Roy A (2001) The algebra of infinite justice. Penguin, New Delhi

Rudolph F (2007) Bewertung des Beitrags von CDM-Projekten zur nachhaltigen Entwicklung seiner Gastländer, Policy paper 3. Wuppertal Institut für Klima, Umwelt, Energie GmbH, Wuppertal

Schneider L (2007) Is the CDM fulfilling its environmental and sustainable development objectives? An evaluation of the CDM and options for improvement. Öko-Institut e.V., Berlin

Smith K (2007) The carbon neutral myth. Offset indulgences for your climate sins. Carbon Trade Watch, Amsterdam

Streck C (2004) New partnerships in global environmental policy: the Clean Development Mechanism. J Environ Dev 13(3):295–322

Strobl T, Zunic F (2008) Wasserbau – Aktuelle Grundlagen, Neue Entwicklungen. Springer, Berlin

Sutter C (2003) Sustainability check-up for CDM projects: how to assess the sustainability of international projects under the Kyoto Protocol. Wissenschaftlicher Verlag, Berlin

Sutter C, Parreño JC (2007) Does the current Clean Development Mechanism (CDM) deliver its sustainable development claim? An analysis of officially registered CDM projects. Clim Chang 84(1):75–90

Tangri N (2003) Waste incineration: a dying technology. Global Alliance for Incinerator Alternatives (GAIA), Quezon City. http://www.no-burn.org/downloads/Waste%20Incineration%20%20A%20Dying%20Technology.pdf. Accessed 10 July 2010

Thakkar H (2009) There is little hope here. India's national action plan on climate change. A civil society view. South Asian Network on Dams, Rivers and People (SANDRP), New Delhi

The Economist (2009) Sin aqua non. Dams are making a comeback. Online edition. http://www.economist.com/node/13349220. Accessed 10 July 2010

Turpin T (2008) Dam. Reaktion Books, London

UNEP (United Nations Environmental Programme) Risø Centre (2010) CDM pipeline overview. http://www.cdmpipeline.org. Accessed 10 July 2010

UNESCO (United Nations Educational, Scientific and Cultural Organization) (2009) The United Nations world water development report 3. Paris

UNFCCC (United Nations Framework Convention on Climate Change) (1992) United Nations Framework Convention on Climate Change. http://unfccc.int/resource/docs/convkp/conveng.pdf. Accessed 10 July 2010
UNFCCC (United Nations Framework Convention on Climate Change) (1998) Kyoto Protocol to the United Nations Framework Convention on Climate Change. http://unfccc.int/resource/docs/convkp/kpeng.pdf. Accessed 10 July 2010
UNFCCC (United Nations Framework Convention on Climate Change) (2007a) Project design document. Allain Duhangan Hydropower Project. http://cdm.unfccc.int/UserManagement/FileStorage/LYZSN7J5RUYNO8DZTC236SF3POYNWK. Accessed 10 July 2010
UNFCCC (United Nations Framework Convention on Climate Change) (2007b) Project design document. Xiaoxi Hydropower Project, China. http://www.netinform.net/KE/files/pdf/PDD_070216.pdf. Accessed 10 July 2010
UNFCCC (United Nations Framework Convention on Climate Change) (2010) Consolidated baseline methodology for grid-connected electricity generation from renewable sources (ACM002). http://cdm.unfccc.int/UserManagement/FileStorage/HGY3TLRFPQVM016WA4I7XCZD92KE5S. Accessed 10 July 2010
WCD (World Commission on Dams) (2000a) Dams and development. A new framework for decision-making. Earthscan, London
WCD (World Commission on Dams) (2000b) Experience with dams in water and energy resource development in the People's Republic of China, Cape Town
Witt U, Moritz F (2008) CDM – saubere Entwicklung und dubiose Geschäfte. In: Altvater E, Brunnengräber A (eds) Ablasshandel gegen Klimawandel? VSA-Verlag, Hamburg, pp 88–106
Wright DV (2007) The Clean Development Mechanism: climate change equity and the south-north divide. VDM Verlag, Saarbrücken
WWF (World Wide Fund for Nature) (2007) Regenwald für Biodiesel? Ökologische Auswirkungen der energetischen Nutzung von Palmöl, Frankfurt am Main

About the Editor/Authors

Ravi Baghel is a postdoctoral researcher examining the production of Himalayan glaciers as objects of environmental knowledge. He conducts this research as a member of the interdisciplinary research group 'Waterscapes' funded by the Cluster of Excellence: Asia and Europe in a Global Context, based at the University of Heidelberg, Germany. He obtained a PhD in Geography from the University of Heidelberg in 2013, supported by a scholarship of the Cluster of Excellence. His doctoral research, carried out as a member of the project 'Large dams: Contested environments between hydro-power and resistance', used a poststructural political ecology approach to investigate the role of expert knowledge in the transformation of Indian rivers through dams and embankments. He received an MA in Social Sciences from the University of Freiburg, Germany, and the University of KwaZulu-Natal, South Africa, in 2008, and a bachelor's degree in Chinese from the Jawaharlal Nehru University, India. He is a fellow of the Earth System Governance Project, and his research interests are in the interaction of knowledge and space, political ecology and South Asian environmental issues.

Kumkum Bhattacharyya studies the interactions between riverine environments and human societies using GIS-based approaches to environmental assessment and management. Dr. Bhattacharyya obtained her PhD degree in Environmental Geography (2000) from the University of Burdwan, India, followed by postdoctoral research at the University of California, Berkeley (2002–2003), where she worked on sediment management in aging reservoirs. She has worked professionally at Washington State Department of Ecology prioritising environmental cleanup activities, contributed to the development of the 'Oregon Explorer Wetland Portal' and helped develop a statewide GIS database of wetlands for Oregon, and she has taught both Geography and GIS at Eastern Michigan University, Saginaw Valley State University in Michigan State and Highline Community College in Washington State. Her book *The Lower Damodar River, India: Understanding the Human Role in Changing Fluvial Environment* was recently published (2011) by Springer. As Research Associate at the School of Natural Resources and Environment,

M. Nüsser (ed.), *Large Dams in Asia*, Advances in Asian Human-Environmental Research, DOI 10.1007/978-94-007-2798-4,
© Springer Science+Business Media Dordrecht 2014

University of Michigan, she is currently involved in several studies in the USA and India exploring river flooding cycles and their relationship to both ecological and social systems.

Nirmalya Choudhury obtained a master's degree in Economics from the Calcutta University in 2002, followed by a postgraduate degree in Forest Management from the Indian Institute of Forest Management in Bhopal in 2005. He then worked with the International Water Management Institute (IWMI) in the IWMI-Tata Water Policy Programme till 2007. During this period he worked on various aspects of institutional reforms in the Indian surface irrigation sector. His work, during this period, has been subsequently published in various national and international journals. Subsequently he worked with Sir Ratan Tata Trust (SRTT) and with Collectives for Integrated Livelihood Initiatives (CInI) on strengthening water-centric livelihoods among the tribal populations in central India. In September 2008 he joined the Deutsches Institut für Entwicklungspolitik (DIE) research team on dams and continued his PhD with the Bundesministerium für Bildung und Forschung (BMBF) IPSWaT scholarship at Technical University of Berlin. His current research work focuses on the social and environmental governance over large water infrastructure projects, particularly the hydropower projects, in India.

Rohan D'Souza is Assistant Professor at the Centre for Studies in Science Policy. He was awarded his PhD from the Centre for Historical Studies at Jawaharlal Nehru University. His research on hydraulic manipulation in the Indian subcontinent was published by Oxford University Press in 2006 as a monograph entitled 'Drowned and Dammed: Colonial Capitalism and Flood Control in Eastern India (1803–1946)'. He has jointly edited a volume entitled *The British Empire and the Natural World: Environmental Encounters in South Asia*, Oxford University Press, 2011. His interests and research publications cover themes in environmental history, conservation, ecological politics, sustainable development and modern technology. He was a postdoctoral fellow at the Agrarian Studies Program (Yale University) and a Ciriacy-Wantrup fellow (University of California, Berkeley). He was also a Senior Research Associate at the Centre for World Environmental History (University of Sussex) and visiting fellow at the Resources Management Asia-Pacific (Australian National University).

Alexander Erlewein is a PhD student at the Geography Department of the South Asia Institute, Heidelberg University, Germany. He studied geography, political science and political science of South Asia and received his master's degree in 2009. His current PhD work analyses the environmental and socioeconomic impacts of hydropower development in Himachal Pradesh, India, with a particular focus on dams which are promoted through the carbon-offsetting scheme Clean Development Mechanism (CDM). His PhD is supported by a scholarship of the research cluster 'Asia and Europe in a Global Context' at Heidelberg University. Further research interests include development studies, climate policy, and water resource management.

Marcus Nüsser is Professor for South Asian Geography with a special focus on human-environment interactions at the South Asia Institute (SAI) of the Universität Heidelberg. He studied Geography, Geology and Development Sociology at the Freie Universität Berlin, leading to a doctoral degree in the Natural Sciences in 1996. In 2003 he completed his habilitation at the University of Bonn. His research focuses on issues of human-environment interaction, environmental dynamics, and development processes, in high mountain regions of South Asia and sub-Saharan Africa. In recognition of his long-standing mountain research, he was nominated to become a member of the Association for Comparative High Mountainous Studies (Arbeitsgemeinschaft für vergleichende Hochgebirgsforschung e.V.). In the Himalayas and the adjoining Karakoram and Hindu Kush, he has conducted numerous field campaigns since 1992 and has published widely in national and international journals. He serves on the editorial boards of the journals *Mountain Research and Development* (since 2003) and *Journal of Mountain Science* (since 2005). Since 2009, he also works as Series Editor for the book series *Advances in Asian Human-Environmental Research.* He has been the recipient of numerous grants for high mountain research.

Miriam Seeger studied Chinese studies and philosophy at Heidelberg University and at Beijing University. In 2012 she received her PhD on large dams and the quest for productive landscapes in China in the twentieth and early twenty-first century from Heidelberg University. From 2008 to 2011 she joined the project 'Large dams: Contested environments between hydro-power and resistance' at the Cluster of Excellence 'Asia and Europe in a Global Context: Shifting Asymmetries in Cultural Flows'. In that time her research was funded by the Cluster of Excellence. She works on environmental politics and history, changing perceptions of nature throughout time and the influence of modernity and development ideas in China.

Hanna Werner received her PhD on Dams and Developmental Debates in twentieth-century India from Humboldt University, Berlin. From 2008 to 2011, she worked as a PhD student at the Cluster of Excellence 'Asia and Europe in a Global Context: Shifting Asymmetries in Cultural Flows', Heidelberg University. She works on debates about modernity, development, governmentality, social movements and the dynamics of knowledge systems in a cross-cultural perspective. She taught at Freie Universität Berlin, Heidelberg University, and University of Chicago and currently is teaching at Humboldt University, Berlin.

Michael J. Wiley is the current Roosevelt Professor of Ecosystem Management at the School of Natural Resources and Environment, University of Michigan. An aquatic ecologist, he has published over 80 peer-reviewed scientific papers on a wide variety of topics including river fish communities, invertebrate biology, land use impacts on water quality, river ecosystem modelling and the eco-hydrology of riparian wetlands. Recent papers explore the impact of climate change on rivers in the Laurentian Great Lakes and farming impacts on stream ecosystems. Since 2004 he and his students have also been working on water resource issues in India and more recently South Korea. His studies in India currently focus on the ecology of lower Ganges and its major tributaries, including the Damodar.

Index

M. Nüsser (ed.), *Large Dams in Asia*, Advances in Asian Human-Environmental Research, DOI 10.1007/978-94-007-2798-4,
© Springer Science+Business Media Dordrecht 2014

MIX
Papier aus verantwortungsvollen Quellen
Paper from responsible sources
FSC® C105338

If you have any concerns about our products,
you can contact us on
ProductSafety@springernature.com

In case Publisher is established outside the EU,
the EU authorized representative is:
Springer Nature Customer Service Center GmbH
Europaplatz 3, 69115 Heidelberg, Germany

Printed by Libri Plureos GmbH
in Hamburg, Germany